Series of the Institute for Climate Protection, Energy and Mobility (IKEM)

In this book series, the Institute for Climate Protection, Energy and Mobility (IKEM) presents its research on climate change mitigation, exploring the intersection of law, economics, politics, and technology. The series offers analyses on all aspects related to achieving a more sustainable society and advancing the transition of energy and mobility systems.

Founded in 2009, IKEM is an interdisciplinary research institute led by its Director, Prof. Dr. Michael Rodi, who conceives this book series together with Dr. Simon Schäfer-Stradowsky. Affiliated with the University of Greifswald, IKEM has held Special Consultative Status with the United Nations Economic and Social Council since 2017.

Jana Karras

Legal Barriers to the Energy Modernisation of Dwellings Occupied by Low-income Tenants and Opportunities to Overcome these Barriers

Case Study of Germany

Springer

Jana Karras
Halle (Saale), Germany

Inauguraldissertation zur Erlangung des akademischen Grades eines Doktors der Rechte der Rechts- und Staatswissenschaftliche Fakultät der Universität Greifswald vorgelegt von Jana Karras 2023

ISSN 2731-3085 ISSN 2731-3093 (electronic)
Series of the Institute for Climate Protection, Energy and Mobility (IKEM)
ISBN 978-3-658-44192-0 ISBN 978-3-658-44193-7 (eBook)
https://doi.org/10.1007/978-3-658-44193-7

This Springer imprint is published by the registered company Springer Fachmedien Wiesbaden GmbH, part of Springer Nature.
The registered company address is: Abraham-Lincoln-Str. 46, 65189 Wiesbaden, Germany

Paper in this product is recyclable.

Acknowledgements

I extend my deepest gratitude to Prof. Dr. Michael Rodi for fostering an academic environment at IKEM research academy that immensely supports young academics. His guidance has been invaluable throughout my doctoral journey, and I am particularly thankful for his role as my supervisor and academic support during this time. I sincerely appreciate my second reviewer Prof. Dr. Schomerus, whose expert guidance and insightful critiques have been indispensable in enhancing the quality of this doctoral research. My heartfelt thanks go to Dr. Aleksandra Novikova, whose support, feedback, and motivation have been pivotal throughout the dissertation writing process. I am equally grateful to the Research Academy team and our academic circle for their invaluable feedback, which enriched my work significantly. A special thank you to Prof. Dr. Rodi and Dr. Simon Schäfer-Stradowsky for the opportunity to publish this book in the book series of the Institute for Climate Protection, Energy and Mobility. Finally, my profound appreciation to Konstantin Karras, whose belief in me and unwavering support over the years have been a constant source of strength and inspiration.

Contents

About the Author

Dr. Jana Karras has been a research associate at IKEM since 2021. She completed her law studies at the Yaroslav Mudryi National Law University in Ukraine, with a focus on international public law. She furthered her education in law, specialising in public law, at the University of Greifswald in Germany. During her legal clerkship, Dr. Jana Karras gained invaluable experience at various institutions, including the Federal Protection Agency and the Luther Law Firm. Her work at IKEM is centered around international climate policy and the energy efficiency of buildings, showcasing her dedication to combining her legal expertise with sustainable energy solutions.

Abbreviations

ALG II	Arbetislosengeld nach SGB II (unemployment benefit in accordance with the Social Code II)
Art.	article
BBSR	Federal Institute for Research on Building, Urban Affairs and Spatial Development
BEHG	Bundesemissionshandelsgesetz (Federal Fuel Emissions Trading Act)
BGB	Bürgerliches Gesetzbuch (Civil Code)
BlnVerf	Verfassung von Berlin (Constitution of Berlin)
BremVerf	Landesverfassung der Freien Hansestadt Bremen (Constitution of Bremen)
BVerfG	Federal Constitutional Court
CEPA	Classification of environmental protection activities
cf.	compare
CFREU	Charter of Fundamental Rights of the European Union
CNY	Chinese yuan renminbi
CO2	carbon dioxide
CRMA	Classification of resource management activities
CSR	Corporate social responsibility
ECHR	European Convention on Human Rights
ECJ	European Court of Justice
ECL	European Climate Law
EED	Directive 2012/27/EU on energy efficiency
e.g.	exempli gratia; for example
EGD	European Green Deal
EnergieStG	Energiesteuergesetz (Energy Tax Act)
EPBD	Directive 2010/31/EU on the energy performance of buildings
EPOV	EU Energy Poverty Observatory
ESR	Regulation (EU) 2018/842 on binding annual greenhouse gas emission reductions by Member States from 2021 to 2030 contributing to climate action to meet commitments under the Paris Agreement and amending Regulation (EU) No 525/2013

ESRa	Energy Transition in Social Space
etc.	et cetera
ETS	Emission Trading System
Et seq.	et sequens; and what follows
EU	European Union
EUR	euro
f.	following
FDP	Free Democratic Party
GCF	Global Climate Forum
GDP	Gross domestic product
GEG	Gebäudeenergiegesetz (Building Energy Act)
GG	Grundgesetz (Basic Law)
GHG	Greenhouse gas emissions
GW	Germanwatch
HeizkostenVO	Heizkostenverordnung (Heating Costs Ordinance)
HGrG	Haushaltsgrundsätzegesetz (Budget Principles Act)
HmbVerf	Verfassung der Freien und Hansestadt Hamburg (Constitution of Hamburg)
i.a.	inter alia; among other things
i.e.	id est; for example
IEE	Fraunhofer Institute for Energy Economics and Energy System Technology
IKEM	Institute for Climate Protection, Energy and Mobility
IPCC	Intergovernmental Panel on Climate Change
IRS	Leibniz Institute for Spatial Social Research
KfW	Kreditanstalt für Wiederaufbau (Reconstruction Loan Bank)
KredAnstWiAG	Gesetz über Kreditanstalt für Wiederaufbau (Law on Reconstruction Loan Bank)
KWG	Gesetz über das Kreditwesen (German Banking Act)
LHO Bln	Berliner Landeshaushaltsordnung (Financial Regulation of Berlin)
MGDD	Manual of Government Deficit and Debt
MHC	Municipal housing company
MietAnpG	Mietrechtsanpassunsggesetz (Act Supplementing the Regulations on the Permissible Amount of Rent at the Beginning of the Tenancy and Adjusting the Regulations on the Modernisation of the Rented Property)
MietenWoG Bln	Gesetz zur Mietenbegrenzung im Wohnungswechsel in Berlin (Law on Rent Limitation in Housing in Berlin)
MietNovG	Mietrechtsnovellierungsgesetz (Act to Control Rent Increases in Tight Housing Markets and to Strengthen the Order Principle in Housing Brokerage)
MietRÄndG	Mietrechtsänderungsgesetz (Act on the Energetic Modernisation of Rented Housing and on the Simplified Enforcement of Eviction Orders)

Mio	million
MsRG	Mietspiegelreformgesetz (Act on the Reform of Rent Index Law)
NECP	Integrated National Energy and Climate Plan
OECD	Organisaton for Economic Co-operation and Development
p.	page
para.	paragraph
pp.	pages
RWI	Renovation Wave
s.	sentence
SAP	Swedish Social Democratic Workers' Party
SEK	Swedish krona
SGB	Sozialgesetzbuch (Social Security Code)
SPD	Social Democratic Party of Germany
Sub-para.	sub-paragraph
TEU	Treaty on the European Union
TF	Tryckfrihetsförordningen (the Freedom of Express Act)
TFEU	Treaty on the Functioning of the European Union
UK	United Kingdom
UNDHR	Universal Declaration of Human Rights
UNFCC	UN Framework Convention on Climate Change
USD	US dollar
VV-WoFGWoBindG	Verwaltungsvorschrift zum Wohnraumförderungs- udn Wohnungsbindungsgesetz (Administrative Regulation on the Housing Promotion and Housing Bond Act)
VwVfG	Verwaltungsverfahrensgesetz (Administrative Procedural Act)
WiStG	Wirtschaftsstrafgesetz (Act on the Further Simplification of Commercial Criminal Law)
WoFG	Gesetz über die soziale Wohnraumförderung (Act on Promoting of Social Housing)
WoGG	Wohngeldgesetz (Housing Allowance Act)

1 Introduction

1.1 Setting the Scene

Human-induced climate change and its consequences are one of the greatest challenges to the global community. Consequently the 196 Parties of the UN Framework Convention on Climate Change (UNFCCC) recognised the urgent need to cap global warming at well below 2 degrees Celsius above pre-industrial levels, aiming for a 1.5-degree Celsius increase. This resolution was solidified at COP 21 in Paris through the Paris Agreement.[1] Germany is party to the UNFCCC and the Paris Agreement.[2] Germany, as a signatory to both the UNFCCC and the Paris Agreement, is obligated to contribute to the global reduction of greenhouse gas (GHG) emissions, as outlined in Article 4 of the Paris Agreement. By ratifying this Agreement, Germany committed to striving for the peak of GHG emissions as soon as possible.

However, effectively addressing the challenges posed by climate change requires more than national efforts alone. The European Union (EU) has set an ambitious objective to become the world's first climate-neutral continent by 2050.[3] The decarbonisation

[1] UNFCCC, Germany Climate Action Plan 2050, p. 2., accessed on 04.03.2021; Calliess/Dross 2020, p. 457.

[2] The Act of Assent to the Paris Agreement of 12 December 2015 has come into force on 04.11.2016, see BGBl. II 2016, Nr. 31, 21.11.2016, p. 1240; Calliess/Dross 2020, p. 457.

[3] The European Green Deal – European Commission'. 2021. 14 July 2021. https://commission.europa.eu/strategy-and-policy/priorities-2019-2024/european-green-deal_en

J. Karras, *Legal Barriers to the Energy Modernisation of Dwellings Occupied by Low-income Tenants and Opportunities to Overcome these Barriers*, Schriftenreihe des Instituts für Klimaschutz, Energie und Mobilität,
https://doi.org/10.1007/978-3-658-44193-7_1

of the building sector plays a crucial role in achieving this ambitious goal, as it is responsible for 36% of all CO_2 emissions in the EU.[4] Decarbonisation of the building sector and improvement of its energy efficiency are important actions to achieve this ambitious goal.[5]

Presently, only 11% of the EU's existing building stock undergoes renovation annually. However, this low renovation rate frequently fails to apply the most modern energy efficiency standards, often neglecting energy performance entirely. Deep renovations, reducing energy consumption by at least 60%, have only been performed on 0.2% of the building stock per year across the EU.[6]

For Germany, decarbonising the residential building sector and improving its energy efficiency is not just crucial, considering the aforementioned challenges; it is also a significant hurdle due to the country's high rate of rented dwellings, which stood at roughly 49% in 2019, one of the highest rates in Europe.[7] This situation highlights the divergence of interests between landlords and tenants in the context of decarbonisation. Landlords prioritise the economic benefits and market value associated with thermal energy efficiency rather than energy conservation itself. High energy performance in properties tends to elevate their sale prices, impacting landlords' rental incomes positively and minimising vacancy risks. However, market imperfections make it challenging for landlords to transfer these investments to tenants, creating what literature terms as the "landlord-tenant dilemma".[8] In other words, landlords hesitate to invest in energy modernisation for rented properties when unsure of the investment's profitability and its impact on their profits. At the same time, tenants are reluctant to invest in energy upgrades for properties they don't own, uncertain about the duration and extent of their benefits.

This complexity escalates when tenants belong to vulnerable social groups with low incomes. They usually have only a limited or no access to the financial assets so that they cannot contribute to the energy modernisation, even if they desire to do so. Consequently, this social group often resides in energy-inefficient housing due to market-driven rent increases. Thus, in 2019, roughly 15% of the German population was at

---

[4] Carbon dioxide (CO_2) is one of the main GHGs besides water vapour, methane (CH_4), nitrous oxide (N_2O) and ozone.

[5] 'New Rules for Greener and Smarter Buildings Will Increase Quality of Life for All Europeans - European Commission'. 2019. 15 April 2019. https://commission.europa.eu/news/new-rules-greener-and-smarter-buildings-will-increase-quality-life-all-europeans-2019-04-15_en.

[6] COM (2020) 662 final.

[7] Statistisches Bundesamt. 2022. 'EU-Vergleich: Überbelastung Durch Wohnkosten 2022'. 2022. https://www.destatis.de/Europa/DE/Thema/Bevoelkerung-Arbeit-Soziales/Soziales-Lebensbedingungen/Wohnkosten.html.

[8] Karras/Novikova 2021, p. 1.

risk of monetary poverty,[9] although this figure decreased in recent years. However, considering the potential aftermath of the Covid-19 pandemic and the energy crisis provoked by Russian unjustified aggression against Ukraine, there is a high risk of this number rising again. Almost every seventh habitant of Germany is, in fact, a low-income person. Around 14% of German households considered monthly housing costs a considerable economic burden in 2019, as they spent over 40% of their income on housing.[10]

Germany, being a welfare state with the strong social legislation ensuring social security, justice, and equal opportunities to its citizens as far as possible,[11] faces an acute issue: energy poverty. Interestingly, this phenomenon remains unaddressed in the German National Energy and Climate Plans (NECPs) under the premise that social assistance covers energy expenses for eligible individuals.[12]

The spectre of energy poverty is likely to grow not only in Germany but across the entire EU due to anticipated increases in energy prices.[13] In European emissions trading, the price of CO_2 certificates has been reaching unprecedented highs, hitting €32 in the mid of December 2020. Factors such as economic crises, Brexit uncertainty, and disruptions in gas deliveries from Norway have contributed to this surge.[14] Furthermore, the national emission trading system is anticipated to push energy prices upward. The recent discourse around escalating electricity prices for households as part of the German energy transition has brought forth a newfound awareness of energy poverty, particularly its disproportionate impact on individuals with low incomes coupled with inadequate housing. The social inequality and, as a result, socio-spatial segregation caused by high energy costs will potentially increase in the future.[15] Russia's unprovoked and unjustified aggression against Ukraine significantly disrupted global energy security, highlighting Germany's vulnerabilities, especially its heavy reliance on Russian gas, consequently leading to rapid increases in energy prices.

---

[9] Statistisches Bundesamt. 2021. 'Armutsgefährdungsschwelle und Armutsgefährdung (monetäre Armut) in Deutschland'. Statistisches Bundesamt. 8 February 2021. https://www.destatis.de/DE/Themen/Gesellschaft-Umwelt/Einkommen-Konsum-Lebensbedingungen/Lebensbedingungen-Armutsgefaehrdung/Tabellen/armutsschwelle-gefaehrdung-silc.html.; Karras/ Novikova 2021, p. 1.

[10] Statistisches Bundesamt. 2022. 'EU-Vergleich: Überbelastung Durch Wohnkosten 2022'. 2022. https://www.destatis.de/Europa/DE/Thema/Bevoelkerung-Arbeit-Soziales/Soziales-Lebensbedingungen/Wohnkosten.html.

[11] Sodan/Ziekow 2014, §10 Rn. 3.

[12] NECP 2020, p. 67; Karras/ Novikova 2021, p. 2.

[13] Bouzarovski, in Großmann/Schaffrin/Smigiel 2017, p. 47f.

[14] FAZ.NET. 2020. 'Europäischer Emissionshandel: Rekord für CO2-Zertifikate', 15 December 2020. https://www.faz.net/aktuell/finanzen/finanzmarkt/europaeischer-emissionshandel-rekord-fuer-co2-zertifikate-17103667.html..
The Russian war in Ukraine has also prompted the world to search for faster energy transition ways.

[15] Schaffrin/Smigiel/Großmann, in: Großmann/Schaffrin/Smigiel 2011, p. 1.

The urgent need for a quick energy transition, encompassing the modernisation of residential infrastructure, dominates political and scientific discussions in Germany, primarily focusing on technical, engineering, and economic aspects.[16] However, the regulative barriers related to these topics should not be underestimated. Identifying and addressing these barriers are crucial steps in defining the scope for political intervention and finding viable solutions to the landlord-tenant dilemma specific to Germany.

Sweden, in contrast, identified the challenge of split incentives early on[17] and thoroughly examined its impact on renovating its building stock. Considering the similarities in non-legal contexts between both countries, which will be expounded upon below, Sweden's approach to addressing the landlord-tenant dilemma has potential interest for Germany. While Sweden has not completely resolved this issue, it has gained recognition for its pioneering initiatives in green financing.[18] This innovative approach could serve as a potential solution for Germany in navigating the landlord-tenant dilemma, especially concerning vulnerable social groups. Considering the aforementioned factors, it should be examined whether there are transferable legal results of the landlord-tenant dilemma from the Swedish legal order into the German one. Should the legal context not lead to the desired outcomes, alternative solutions beyond the legal context will also be examined.

1.2 Theoretical and Practical Contribution

The thesis is part of the "Energy Transition in Social Space" (ESRa) project,[19] collaborative effort involving the Global Climate Forum (GCF), the Fraunhofer Institute for Energy Economics and Energy System Technology (IEE), Germanwatch (GW), the Institute for Climate Protection, Energy, and Mobility (IKEM), and the Leibniz Institute for Spatial Social Research (IRS). Additionally, practical support from various stakeholders in politics, business, science, and civil society was instrumental.[20]

Within the ESRa project, the focus was on identifying key factors that differentiate between advantaged and disadvantaged social spaces concerning energy transition. The subsequent step involved exploring methods to positively influence these factors for sustainable regional development. To accomplish this, two comparative case studies were undertaken, engaging citizens and relevant stakeholders. One study was conducted in Berlin, a bustling urban centre reaping the benefits of globalisation. The

[16] Ibid, p. 4.

[17] Boverket 2013.

[18] See Vanhuyse/Chan/Gill 2020, p. 3 f.; Climate Bond Initiative 2018, p. 6.

[19] Energiewende im sozialen Raum'. n.d. 29.08.2022. https://esra-projekt.de

[20] 'Energiewende im sozialen Raum. Projektteam'. n.d. 29.08.2022. https://esra-projekt.de/das-projektteam

second case study took place in the Spree-Neiße district within the Lusatia region of Brandenburg,[21] representing a rural area undergoing structural changes due to factors such as coal mining, alongside extensive agricultural landscapes and tourist attractions like the Spreewald. This region, possibly more vulnerable to the impacts of globalisation and population decline, was examined. One key focus of the comparative studies was the investigation of regulatory and financial barriers associated with enhancing thermal efficiency in rented dwellings across Germany. The quantitative and qualitative data obtained from these studies are applicable beyond the specific regions studied, given the high prevalence of rented dwellings throughout Germany.[22]

When analysing the diverse impacts of the energy transformation on various regions and demographic segments, several central questions were explored:

- Why is there a growing divide between urban and rural areas in the discourse surrounding energy transition?
- How are the pros and cons of this transformation distributed across these social landscapes?
- Do rural areas suffer as losers in this transformation, or do they also stand to gain opportunities for regional development?
- What strategies exist to navigate the socio-spatial challenges presented by the energy transition?

The project unfolded across three distinct phases: initially, an empirical survey scrutinised the current regional situations (phase 1); subsequently, an analysis probed the underlying socio-economic factors and their interconnectedness (phase 2); finally, the project identified potential future trajectories and the necessary measures for their realisation (phase 3). This study formulated and scrutinised various energy transition scenarios: best-case, worst-case, and the most probable scenarios. Each scenario underwent scrutiny through a legal and financial framework, assessing their applicability to different aspects of the energy transition.[23]

In crafting the framework, insights were drawn from other countries' experiences with energy transition projects. Sweden held particular interest due to its municipalities' success in financing energy transition projects via green bonds, contrasting with Germany's more cautious approach to innovative financial instruments. While the general applicability of municipal green bonds in Germany was assessed in the project, specific research

[21] 'Energiewende im sozialen Raum'. n.d. 31.08.2022. https://esra-projekt.de/das-projekt-0

[22] Statista Research Department. 2022. 'Bewohnte Mietwohnungen in Deutschland nach Bundesländern'. Statista. 20 January 2022. https://de.statista.com/statistik/daten/studie/258643/umfrage/bewohnte-mietwohnungen-in-deutschland-nach-bundeslaendern/

[23] ESRa, AP4 2021.

delved into their suitability for funding the energy modernisation of rented dwellings occupied by low-income tenants.

The findings of this thesis significantly contribute to the academic discourse on energy poverty in Germany. It sheds light on an existing problem that the German government has yet to address. Furthermore, it conducts an analysis of the current state of the landlord-tenant dilemma, with a particular focus on vulnerable households. Given the trajectory of increasing energy prices, the proportion of this social group facing challenges may also rise. Additionally, this thesis proposes an innovative solution within the German legal framework for the landlord-tenant dilemma. At its core, the proposed solution does not involve redistributing the modernisation costs within the landlord-tenant relationship but rather advocates for financing energy modernisation through external sources beyond this contractual relationship.

References

Boverket. 2013. 'Analys av delade incitament för energieffektivisering'. 2013:32. https://www.boverket.se/sv/om-boverket/publicerat-av-boverket/publikationer/2013/analys-av-delade-incitament-for-energieffektivisering/.

'Bundesgesetzblatt BGBl. Online-Archiv 1949–2022 | Bundesanzeiger Verlag'. 2016. 21 November 2016. https://www.bgbl.de/xaver/bgbl/start.xav?start=%2F%2F*%5B%40attr_id%3D%27bgbl216031.pdf%27%5D#__bgbl__%2F%2F*%5B%40attr_id%3D%27bgbl216031.pdf%27%5D__1700861902664.

Bundesministerium für Umwelt, Naturschutz und Reaktorsicherheit. 2016. 'Climate Action Plan 2050. Principles and Goals of the German Government's Climate Policy'. https://www.bmu.de/fileadmin/Daten_BMU/Pools/Broschueren/klimaschutzplan_2050_en_bf.pdf.

Climate Bond Initiative. 2018. 'The Green Bond Market in the Nordics 2018'. https://www.climatebonds.net/files/files/CBI-The%20Green%20Bond%20Market%20in%20the%20Nordics.pdf.

Communication from the Commission to the European Parliament, the Council, the European Economic and Social Committee and the Committee of the Regions a Renovation Wave for Europe – Greening Our Buildings, Creating Jobs, Improving Lives. 2020. https://eur-lex.europa.eu/legal-content/EN/TXT/?uri=CELEX%3A52020DC0662.

FAZ.NET. 2020. 'Europäischer Emissionshandel: Rekord für CO_2-Zertifikate', 15 December 2020. https://www.faz.net/aktuell/finanzen/finanzmarkt/europaeischer-emissionshandel-rekord-fuer-co2-zertifikate-17103667.html.

Federal Environment Ministry. n. d. 'Climate Action Plan 2050 (EN)- BMUV - Publication'. Bmuv.De. Accessed 24 November 2023. https://www.bmuv.de/PU395-1.

Grossmann, Katrin, André Schaffrin, and Christian Smigiel, eds. 2011. *Energie und soziale Ungleichheit: zur gesellschaftlichen Dimension der Energiewende in Deutschland und Europa.* Wiesbaden: Springer VS.

Karras, Jana, and Aleksandra Novikova. 2021. 'Germany. Legal Barriers to the Decarbonisation of Dwellings Occupied by Low-Income Tenants and Financial Incentives to Overcome These Barriers'. IKEM Climate Finance Week. https://www.ikem.de/wp-content/uploads/2021/05/Brief_5_Legal-barriers-to-the-decarbonisation-of-dwellings.pdf.

'New Rules for Greener and Smarter Buildings Will Increase Quality of Life for All Europeans – European Commission'. 2019. 15 April 2019. https://commission.europa.eu/news/new-rules-greener-and-smarter-buildings-will-increase-quality-life-all-europeans-2019-04-15_en.

Novikova, Aleksandra, Tim Langenhorst, Jana Karras, Peter Ulrich, and Ulrike Damerau. n. d. 'Synopsis Der Handlungsoptionen, Energiewende Im Sozialen Raum'. Produkt No. 4.1. https://esra-projekt.de/sites/default/files/2021-08/esra_produkt_4.1_statusquo_allianzen_und_finanzierung_.pdf.

Sodan, Helge, and Jan Ziekow. 2014. *Grundkurs Öffentliches Recht: Staats- und Verwaltungsrecht.* 6., Wesentlich überarbeitete Auflage. München: C. H. Beck.

Statista Research Department. 2022. 'Bewohnte Mietwohnungen in Deutschland nach Bundesländern'. Statista. 20 January 2022. https://de.statista.com/statistik/daten/studie/258643/umfrage/bewohnte-mietwohnungen-in-deutschland-nach-bundeslaendern/.

Statistisches Bundesamt. 2021. 'Armutsgefährdungsschwelle und Armutsgefährdung (monetäre Armut) in Deutschland'. Statistisches Bundesamt. 8 February 2021. https://www.destatis.de/DE/Themen/Gesellschaft-Umwelt/Einkommen-Konsum-Lebensbedingungen/Lebensbedingungen-Armutsgefaehrdung/Tabellen/armutsschwelle-gefaehrdung-silc.html.

Statistisches Bundesamt. 2022. 'EU-Vergleich: Überbelastung Durch Wohnkosten 2022'. 2022. https://www.destatis.de/Europa/DE/Thema/Bevoelkerung-Arbeit-Soziales/Soziales-Lebensbedingungen/Wohnkosten.html.

'The European Green Deal – European Commission'. 2021. 14 July 2021. https://commission.europa.eu/strategy-and-policy/priorities-2019-2024/european-green-deal_en.

Vanhuyse, Fedra, and Stefanie Chan. 2020. 'Bonds beyond Green', March. https://www.sei.org/publications/bonds-beyond-green/.

Calliess, Christian, and Miriam Dross. 2020. 'Umwelt- Und Klimaschutz Als Integraler Bestandteil Der Wirtschaftspolitik Überlegungen Im Lichte von European Green Deal Und Corona-Recovery Plan'. ZUR, no. 9: 456–64.

'UNFCCC Climate Action Plan 2050'. 04 March 2021. https://unfccc.int/files/focus/application/pdf/161114_climate_action_plan_2050.pdf.

Research Methodology 2

2.1 Multi-Methodological Approach

A multi-methodological approach to the following research seems necessary as the researched question is very complex and refers not only to different branches of law and legal systems but even to different disciplines. There are numerous reasons for using several methods while researching legal barriers to the decarbonisation of dwellings occupied by low-income households. Firstly, applying only one method e.g., the method of comparative law, could result in merely presenting the status quo of legal regulations in the compared countries. If only the methods of legal interpretation are used, there is a risk of misunderstanding the legal norms of other legal systems. Solely relying on economic methods would pose difficulties in devising a problem solution, particularly without considering the different backgrounds of those countries, which the conceptual method of comparative law does. Speaking about the decarbonisation of multi-residential housing, one should not overlook that not everything technically (and legally) feasible is economically justifiable and thus attractive to an investor.[1] Therefore, economic methods are of high importance. Nevertheless, the best practical results can be achieved by combining several methods.

The aforementioned considerations have influenced the structure of this thesis. Firstly, the applied methods are described, and their necessity for the following research is explained. Then, evaluation criteria, as well as limitations, are set up. Later, the current state of research is explained: what is the landlord-dilemma, what problems are the

[1] Stock in: Härtel/ Pielow 2014, p. 139.

J. Karras, *Legal Barriers to the Energy Modernisation of Dwellings Occupied by Low-income Tenants and Opportunities to Overcome these Barriers*, Schriftenreihe des Instituts für Klimaschutz, Energie und Mobilität,
https://doi.org/10.1007/978-3-658-44193-7_2

low-income households dealing with in the context of energy efficiency and how does the landlord-tenant dilemma overlap with low-income households.

Subsequently, the European legal framework for the context of this thesis is analysed as it directly impacts the legal systems of the compared countries and certain aspects related to energy efficiency are harmonised at the EU level. Following this, both the non-legal and legal contexts of compared countries, namely Germany and Sweden, are scrutinised in light of the research questions formulated in the methodological part of this thesis. Subsequent to this comparison, the results are outlined. Lastly, taking into account the aforementioned elements, potential solutions for the German context are developed. In doing so, proposed solutions from existing literature as well as best practices will be critically assessed, particularly regarding their legal and economic viability.

2.2 Legal Interpretation

As this thesis focuses on a case study of Germany, one of the primary methods employed is the method of legal interpretation, specifically concerning German national and regional laws. This legal instrument encompasses four approaches: the wording of the law, grammatical analysis, historical context, and systematic interpretation. Employing these methods in combination aids in comprehending the essence of the law from diverse perspectives. The method of legal interpretation becomes necessary when addressing legal issues that lack clarity, where the norm cannot be unequivocally understood, and thus, the posed question cannot be directly answered.[2] Therefore, this legal method is highly important, especially for Sect. 5.2.2. of this thesis (German legal framework).

However, the four classic methods of interpretation do not entirely align with Savigny's approach.[3] He distinguished between the grammatical, logical, historical, and systematic elements within the legal interpretation. t is noteworthy that the logical element in Savigny's context did not correspond to the teleological element commonly referenced today, but rather focused on the structure of thought, i.e., the "spiritual content" of the law.[4]

The starting point for any interpretation is the wording of the law.[5] Laws are not unchangeable. The research interest of jurisprudence is, therefore, less directed towards the legal texts, which may stay unchanged, but rather towards the content of the law.[6] For instance, when examining the MietenWoG Bln, understanding the scope of application of this law necessitates knowledge of what constitutes rental housing. The grammatical

[2] Sodan/Ziekow 2014, §2 Rn. 1.

[3] Savigny 1840, p. 213 f.

[4] Kischel 2015, p. 426; Sodan/Ziekow, §2 Rn. 4.

[5] Kischel 2015, p. 427.

[6] Pawlowski 1999, p. 16.

method seeks the customary sense of the word, where in the context of rental housing, it typically refers to the living space rented to tenants. However, the scope of the wording interpretation is limited here, as the legislator has outlined in §1 MietWoG instances when this law does not apply to specific rental housing groups. This is where the systematic method of interpretation becomes relevant. Systematic interpretation involves reading the provision in conjunction with other provisions within the same law, though it is not always mandatory.[7] For example, the aforementioned §1 MietWoG Bln demonstrates systematic interpretation by excluding certain living spaces from the law's scope. Therefore, in interpreting rental housing, it should be understood as living spaces rented to tenants, excluding those specified in §1 MietWoG Bln.

Another crucial method of legal interpretation is the teleological interpretation. This approach aims to interpret provisions in a manner that best serves their purpose. The purpose can either be explicitly intended by the legislature or inferred from a comprehensive, objective understanding of the law as a whole.[8] In understanding the example provided in the MietWoG, it is essential to consider the official documents (Drucksache 18/2347 des Abgeordnetenhaus Berlins) to ascertain the law's purpose.[9] According to these documents, the law's purpose is to prevent further rent increases in Berlin. The legislature felt compelled to intervene in the market development and enact public law measures to restrict rents in Berlin, ensuring the affordability of living spaces until the housing market situation stabilises. This knowledge becomes useful when dealing with situations not explicitly covered by the law, prompting those applying the provisions of MietWoG Bln to question whether the case aligns with the law's purpose and if it makes sense to apply its provisions in that scenario.

Finally, the historical method of legal interpretation is an essential instrument for lawyers and legal researchers. Not only the specific history of the law's origins and the will of the historical legislator are scrutinised with this method, but also the bigger intellectual-historical context in which the norm has been enacted.[10] Examining the entire historical development in the realm of rental properties over the last decade could shed light

---

[7] Kischel 2015, p. 427.

[8] Ibid, p. 428.

[9] The MietWoG Bln has been declared unconstitutional by the Federal Constitutional Court as its regulations on rent levels for freely financed housing that can be offered on the open housing market (unrestricted housing) fall under concurrent legislative competence. The Länder are only authorised to legislate as long as and insofar as the Federation has not made final use of its legislative competence (Art. 70, Art. 72 (1) GG). Since the federal legislature has conclusively regulated rent law in Sections 556 to 561 of the German Civil Code (BGB), there is no room for the legislative power of the Länder due to the blocking effect of federal law. Since the core of the MietenWoG Bln also regulates the rent level for unrestricted housing, it is void as a whole, BVerfG, Beschluss des Zweiten Senats vom 25. März 2021 - 2 BvF 1/20 -, Although this law is invalid, it will still be cited as an example for legal interpretation.

[10] Kischel 2015, p. 429.

on interpreting certain provisions of the MietenWoG Bln, particularly those relating to social hardship.

Differentiating between the various methods of interpretation can often pose challenges. For example, analysing the legal norm through the historical perspective can serve as both a historical and a teleological interpretation. The terminology and associated meanings are correspondingly diverse. For example, historical interpretation is sometimes narrowly linked only to the specific history of a norm's origins, also referred to as genetic or historical-genetic interpretation. There are notable similarities between systematic-logical interpretation and objective-teleological interpretation or between objective-teleological and subjective-teleological approaches. While the specifics of these combinations are debated,[11] they hold no direct practical impact on this research and will not be extensively discussed.

Further interpretation instruments should also be mentioned, namely, analogy, conclusion *a majore ad minus*, reverse conclusion and teleological reduction. However, these instruments are typically viewed more as judicial developments within the law rather than methods of legal interpretation.[12] Nevertheless, they might be used in this research when analysing the German legal framework.

The barriers within the existing legal framework concerning the energy transition in the housing sector will be identified and examined through the aforementioned interpretation methods. Based on the research outcomes, proposals for further amendments and adjustments to the legal framework will be formulated. These suggested solutions could potentially be relevant to other EU member states.

2.3 Comparative Law Analysis

Germany, as a federal republic comprising sixteen federal states (*Länder*), grants each its own competencies. Consequently, German lawyers and legal researchers routinely engage with comparative legal methods, albeit not always consciously. While comparing the legal framework of two German *Länder*, the comparative method of law cannot be neglected. This method is also significant when it should be decided whether the experience of Berlin and Brandenburg applies to other German regions and—even more—to other countries. Furthermore, this thesis aims to examine the best practices of other countries, especially Sweden, to find their solutions to the researched problems. Hence, it is crucial to employ this method accurately to ascertain the relevance of these practices for Germany, as well as whether the German experience and formulated solutions can be applied to Sweden and other countries.

[11] Ibid, p. 429f.

[12] Ibid, p. 430.

Comparative disciplines delve into the similarities or differences among various cultural or social phenomena. Therefore, comparative law can be considered a distinct legal subject within the broader field of comparative disciplines.[13] iven this thesis' objective to compare different legal systems to understand how Sweden addresses the researched problem, a comparative law analysis is necessary. Comparative law focuses on comparing diverse national legal systems. It is sometimes labelled as macro-comparison (*Makrovergleichung*), while micro-comparison (*Mikrovergleichung*) delves into specific legal institutions or issues, such as the specific rulings that govern particular issues or conflicts of interest across different legal systems. Within this form of comparative analysis, the emphasis lies on general approaches to handling legal substance, rather than specific problems and their solutions.[14]

Schneider outlines different forms of comparative law analysis: dogmatic foreign law studies, functional comparative law, and integrative comparative law. Dogmatic foreign law studies often serve as the initial step in legal comparison, aiming to either propose solutions for specific legal problems (micro-comparison) or analyse entire legal systems or legal domains (macro-comparison). These studies can be further enriched by a cultural-scientific, especially a legal-historical approach.[15] However, *Kischel* argues against the existence of "foreign law studies", asserting that the term merely serves to distinguish what comparative law is *not*. Only legal professionals trained in the respective foreign legal systems possess comprehensive knowledge of those systems. Individuals versed in multiple legal systems are more the exception than the norm.[16] The debate on the existence of foreign law studies remains ongoing.[17]

The next stage in legal comparison, as outlined by *Schneider*, involves functional comparative law, which seeks functional equivalents of common legal solutions in two or more legal systems.[18] Functional comparative law focuses on issues and creates the necessary *tertium comparationis*, which serves as the point of comparison.[19] As this thesis aims to find legal barriers to improving thermal efficiency in low-income tenant-occupied dwellings, the form of comparison used for this research is rather micro-comparison. The contextual approach is used as a contribution to gain new perspectives on the studied topic. Organising and systematising material about foreign legal systems is essential for readers' comprehension. However, it is crucial to acknowledge that using German legal terms like "*Verwaltungsakt*", for instance, might evoke mistaken

[13] Jansen in: Reimann/ Zimmermann 2006, p. 306.

[14] Zweigert/ Kötz 1996, p. 4f.

[15] Schneider 2007, p. 30.

[16] Kischel 2015, p. 5f.

[17] Ibid.

[18] Schneider 2007, p. 30f.

[19] Ibid.

associations influenced by the researcher's legal system values. Thus, employing original legal terms rather than translations at times may mitigate potential misunderstandings.

Zweigert and *Kötz* considered functionality as the basic principle of comparative law.[20] *Zweigert* proposed a comparative law definition: when deciding complex cases, the judge may use existing and broadly used German four types of interpretation and a comparative-legal method. This approach provides them with an overview of how similar cases are adjudicated in other countries.[21] According to *Zweigert* and *Kötz*, establishing a precise method in advance is impossible. They consider the methodological approach as adopting a method as a hypothesis and testing its effectiveness and practicality based on the results obtained from its application.[22]

The German philosopher *M. Solomon* initially proposed a functional comparison. Later *Ernst Rabel* and his followers supported this approach.[23] As per the functionality principle, only laws serving the same function can be compared. Furthermore, functionalists argue that the function of a rule serves as the common determinant, known as the *tertium comparationis*, enabling comparison.[24] Therefore, according to this perspective, every society resolves similar legal issues differently. Consequently, any initial query concerning the comparison of diverse legal systems must be made functional, devoid of concepts from one's legal system. This helps avoid the potential danger of seeking a solution that merely mirrors a solution within its own legal framework.

The functional method of comparative law as faced criticism in literature for various reasons, often deemed impractical[25]. However, it proved beneficial in Central and Eastern European countries post the collapse of the Soviet system, where lawmakers needed to reconstruct their legal systems. By comparing the experiences of other European countries,[26] they found solutions aligning with their legal traditions.[27] Similarly, comparing Sweden's handling of climate financing in general, and the landlord-tenant dilemma specifically, could offer insights into solving Germany's researched issue.

Before assessing the adaptability of a foreign solution for Germany, *Zweigert* and *Kötz* emphasise answering two crucial questions: whether the solution proved satisfactory in its original country and whether it can function effectively in the adopting

[20] Zweigert/Kötz 1998, p. 33f.

[21] Saidov 2003, p. 44.

[22] Zweigert, Kötz 1998, p. 33.

[23] Saidov 2003, p. 44f.

[24] Brand 2007, p. 409.

[25] Piek 2013, p. 62; Kischel 2015, p. 95 ff.

[26] However, not only European experience, as Ukraine, for example, has transposed some national criminal procedural institutions, which by its legal nature belong to the Anglo-American justice system (see Tolochko 2014, p. 272 f.).

[27] Zweigert/Kötz, 1998, p. 17.

country.[28] Modifications to the solution might be necessary due to differences in authorities' powers, the broader social context, or even the societal understanding of legal norms.[29] Given the disparities between Swedish and German legal systems, any transposition of Swedish legal experience into the German system (and vice versa) demands careful consideration.

Moreover, when comparing different legal systems, one must acknowledge potential limitations or constraints on the outcomes. here are instances where a legislator enacts a law for purely symbolic reasons, such as the introduction of the federal Tenancy Law Adjustment Act (*Mietrechtsanpassungsgesetz*) in 2019. The legislator tried to address the rent increase: He recognised, however, that it might harm effective energy modernisation.[30] This law was criticised for being symbolic, lacking substantial change in effect.[31] While it is feasible to compare this law with Swedish rental law, it is impractical to transpose this experience into the Swedish legal order.

In addition, certain legal norms that effectively address specific problems may be absent within a particular system. Norms and doctrines existing within a society are intricately linked to its social life, serving specific functions. Notably, specific legal institutions may simultaneously serve diverse functions, and their historical contextualisation bears substantial significance.[32] Finally, the presumption that legal systems universally resolve practical questions in a similar manner could be extended to seemingly "apolitical" legal subjects unaffected by moral viewpoints or values.[33]

Every comparative law research begins with formulation of a working hypothesis, and this dissertation adheres to this established practice. The working hypothesis here asserts that low-income tenants in Germany have legal barriers on their way to thermal efficiency. Given Sweden's reputation for practical solutions in climate financing, it is likely that the country has also addressed solutions to the landlord-tenant dilemma and its associated barriers. Consequently, Germany is greatly interested in Sweden's legal experiences in this sphere, warranting a comprehensive comparative analysis of both legal systems.

Functional comparison is based on three central premises:[34]

- Firstly, law serves as a tool to guide human behaviour and responds to societal needs and interests, hence legal norms are the product of the "problem-solution-approach".

---

[28] Ibid.

[29] Ibid.

[30] Bundestag, Drucksache 19/4672, p. 24.

[31] Artz/Börtinghaus 2019, p. 25.

[32] Brand 2007, pp. 415–417.

[33] Brand 2007, p. 418f.; Zweigert/Kötz 1996, p. 12.

[34] Brand 2007, p. 409f.

- Secondly, the issues addressed by legal norms are of sociological concern, and different legal systems aim to resolve similar or identical problems.
- Thirdly, legal systems tend to resolve practical questions in the same way.[35]

Zweigert and *Kötz* described this presumption of similar practical results as a "*praesumptio similitudinis*"[36]. It assists in initially identifying the areas of foreign law likely to reveal similarities and alternatives. It also aids in evaluating whether the comparative outcome aligns with or diverges significantly from similar practical results. However, *Zweigert* and *Kötz* aution that significant deviations or opposing outcomes should prompt a re-evaluation to ensure the original question was functionally formulated, and the research scope sufficiently broadened.[37] Nevertheless, this presumption is subject to limitations, which will further discussed below.

It is essential to understand the functions of the functional method in comparative law as it helps later evaluate the results. *Michaels* has described seven functions of the functional method in comparative law:

- the epistemological function of understanding legal rules and institutions,
- the comparative function of achieving comparability,
- the presumptive function of emphasising similarity,
- the formalising function of system building,
- the evaluative function of deterring the better law,
- the universalising function of preparing legal unification, and last but not least
- the critical function of providing tools for the critique of law.[38]

This method assists in determining legal institutions across different legal orders related to the research question, acknowledging their non-obligatory nature. Different legal systems may employ distinct instruments to address similar issues. Thus, the functional method aids in formulating research questions to achieve comparability and identify similarities. It facilitates a systematic approach for conducting comparative analyses, working towards comprehending and evaluating the comparison's outcomes. Fundamentally, the functional method perceives legal institutions not merely as doctrinal constructs but as societal responses to problems, mandating their observation and comparison within the broader legal system.[39]

Applied to this study, while comparing legal norms of diverse legal orders, the focus lies not solely on the letter of the law but also on the "law in action", encapsulating

[35] Ibid.

[36] Zweigert/ Kötz 1998, p. 40.

[37] Ibid.

[38] Michaels in: Reimann/ Zimmermann 2006, pp. 362–363.

[39] Ibid, pp. 364–366.

the spirit of the law. Furthermore, owing to the comparative function of functionalism, certain abstract issues are presumed universal.[40] *Michaels* asserts the survival needs of individuals as an example of such an abstract universal problem. However, considering climate change's negative impact, it could similarly be regarded as a universal issue. On a detailed level, both Sweden and Germany confront the effects of climate change, aiming to reduce CO_2 emissions as EU member states. They share common challenges such as unrenewed building stock, a proportion of rented dwellings, and populations experiencing low income. Consequently, both nations likely grapple with the landlord-tenant dilemma and energy poverty in comparable ways.

According to *Michaels*, the universality of problems prompts an inquiry into differences and similarities among different societies. It is assumed that diverse societies have similar needs and, consequently, develop functionally equivalent institutions to meet these needs, essential for societal survival. From a functional perspective, the absence of a functional equivalent in a foreign legal order should prompt an examination of whether the original question was formulated purely on functional grounds. Additionally, it is advisable to assess whether the research scope was broad enough.[41]

Applying this thesis to this premise implies that Sweden faces a comparable, if not identical, issue—improving thermal efficiency in dwellings occupied by low-income tenants. The foundational framework is established by EU regulations, such as Regulation (EU) 2018/842, which mandates similar greenhouse gas emission reduction targets by 2030: 38% for Germany and 40% for Sweden. Consequently, both nations operate under similar EU-mandated goals but retain autonomy in transposing these regulations into national law. Legal norms governing issues related to this matter are thus shaped by societal needs, leading to a presumption that these countries may approach the issue similarly.

Kischel emphasises that selecting legal systems, formulating functionally designed questions, and iteratively refining these inquiries help create country reports that objectively reflect the legal landscape, devoid of evaluation based on respective countries' systematic or terminological perspectives. Therefore, identified solutions should be juxtaposed and analysed for similarities and differences, disregarding national dogma. Following the identification of differences, a comprehensive analysis of their legal and other underlying reasons is required. This may require in-depth and comprehensive investigations that extend beyond the realms of legal dogmatics. Ultimately, at the conclusion of the inquiry, an assessment of the results should be undertaken, potentially highlighting superior or inferior solutions, yet this evaluation remains non-mandatory.[42]

The research questions addressed in this dissertation are assumed to fall under public law norms. Public law does not provide the same voluntary frameworks as private law

---

[40] Ibid p. 368.

[41] Ibid, p. 369.

[42] Kischel 2015, p. 93f.

does. It does not provide remedies when the individuals exceed the bounds of the acceptable use of personal power.[43] In public law, the state imposes sanctions upon violation of established norms. Individuals have instruments for legal protection only if the state misuses sanctions and violates their rights. However, civil law is also of high importance here, as tenancy law is private by its legal nature and regulates many aspects of the landlord-tenant issue studied. Moreover, certain aspects, such as the issuance of green bonds in Germany, may remain unregulated by legal norms.

Bell contends that a correction in the comparison of functions within public law is essential. Relying solely on comparisons based on similarities may be inherently misleading. He emphasises the importance of situating comparative activities within the institutional framework of law. While a broadly generic function may exist across multiple legal systems, a nuanced examination within distinct institutional settings presents considerable challenges for comparisons.[44] The distinctiveness of public law in individual countries revolves around institutions that have a different design but are also evoked by history and social problems such as religious and linguistic diversity.[45]

According to *Bell*, the essence of Public Law Comparison summarises the premise that legal activities encompass both functional tasks and their procedural execution. He also emphasises the importance of discovering the external environment and the power of history and routine while comparing public law of different countries.[46] encompasses inputs from national and supranational legislators that significantly impact functional dynamics.[47] The supra-national influence provides an additional reason for considering public law not in isolation but in the global context.[48] Looking at the historical background helps to understand the development of legal and governmental institutions and provides points of reference for issues with which public law has to deal.[49]

However, even if presumed, similar problems faced by chosen countries and their similar approaches may still be influenced by various factors. *Kischel* proposes the contextual method for legal comparison, considering context as a core concept for both individual and typified overall comparisons. For an individual comparison, considering the legal and non-legal environment surrounding the relevant legal norm is essential. Elements relevant to individual comparison should be derived from practical experience. To gain this practical experience regarding foreign legal orders, the hermeneutic approach should be employed to develop a foundational understanding of the

[43] Bell in Harding/Örücü 2002, p. 236.
[44] Ibid, p. 236.
[45] Ibid, p. 240f.
[46] Ibid, p. 236.
[47] Ibid, p. 236f.
[48] Ibid, p. 240.
[49] Ibid, p. 241.

retrospective legal system. Using the legal knowledge and experiences of other comparative lawyers makes sense to comprehend foreign law and its concepts.[50]

Kischel indicates the following elements that may be relevant for the context:

- Historical aspects,
- Non-legal elements,
- Variety of individual elements.[51]

However, this list is non-exhaustive and may vary depending on the specific comparison.

Historical aspects are relevant only to the extent that they influence the present context. Moreover, empirical studies, statistics, anthropological studies on everyday culture, and attitudes of the retrospective nation can be used for contextual legal comparison. The comparative law expert is not limited by empiricism but may also rely on observations or experiences of other comparative law experts or foreign lawyers. When using sociological studies, the comparative law expert should view them with a legal perspective.[52] All in all, the context encompass diverse elements such as the form and content of legal education, styles of judicial decisions, judiciary self-conceptions, the role of legal practitioners, and the level of distrust towards the state or its specific parts.[53]

Kischel distinguishes between several contexts with their sub-contexts. As the chosen legal orders belong to the Continental-European context, they share similarities that render them understandable even without expertise[54]. Nonetheless, these countries belong to different sub-contexts, implying that approaches to the same issue might be housed in unexpected law institutions.

For instance, the Scandinavian context, which includes Sweden, exhibits certain distinct attitudes. There is no codification in the sense of BGB.[55] Sweden has a relatively small number of inhabitants and correspondingly a low number of lawyers and scholars. Thus, the volume of legal literature there is significantly lower than in Germany.[56] Moreover, Scandinavian pragmatism, termed "*lagom*" in Swedish, implies sufficiency, adequacy, or the notion of "just right". There is a low level of legal actions in Scandinavian countries because of the high degree of social control, a still recognisable society pattern of peasants, as well as egalitarian features of the societies. In Sweden, legal challenges against administrative decisions are often prohibited by law.[57]

[50] Kischel 2015, p. 239.

[51] Ibid, p. 241f.

[52] Ibid, p. 241.

[53] Ibid, p. 242.

[54] ibid, p. 389.

[55] Ibid, p. 599f.

[56] Ibid, p. 607.

[57] Ibid, p. 609.

However, the country has a Parliamentary Ombudsman (*Riksdagens ombudsmän*, or *Justitieombudsmannen*) tasked with supervising the executive.[58]

Another typical feature of the Scandinavian context is the principle of publicity. In Sweden, every citizen possesses the right to examine official documents unless they involve matters like state security protection, foreign relations, crime prevention, specific economic interests, or the inviolability of private life (Chapter 2 §1, §2 Tryckfrihetsförordningen (TF, The Freedom of Expression Act). This principle enables inspection not only of all court and authority acts, including government and parliamentary communications, but also encompasses correspondence between authorities and civil servants and all decisions made within the examined case, even before finalisation.[59]

Another characteristic of the Scandinavian context is the lesser significance attributed to constitutional law alongside the backdrop of the welfare state.[60] The state is not seen as a potential threat to the freedoms of its citizens but rather as a guarantor of their well-being. Consequently, no specific constraints by higher-ranking guideline are deemed necessary.[61] Last but not least, the Scandinavian context can be characterised by weak judicial control. Despite the absence of dedicated constitutional courts, all courts are empowered to directly apply the Constitution, albeit without *erga omnes* effect.[62]

At first glance, the Swedish context may appear fundamentally different from the German one. However, as previously noted, absolute statements should be approached cautiously. Their EU membership suggests some level of similarity. Both countries have a long history of relationships that has been lasted for centuries. Sweden had a close relationship with the Hanseatic League. In 1278, Greifswald residents established a herring trade centre near Malmö in Skanör, later expanding to Bornholm, becoming a vital hub for Danish-Swedish-Scandian trade. Through the powerful influence of the Hanseatic League, Stockholm was, for a long period, a half-German town. The German merchants and manufacturers who migrated to Stockholm provided more than half of the municipal tax revenue in the 14th century. The influence of Low German merchants on modern Scandinavian speech is still apparent. Rostock University, established in 1419, held considerable sway over the Northern European region, later shared by Greifswald University, established in 1456. *Samuel von Pufendorf*, a renowned German jurist, lectured at Lund University. Additionally, a Swedish Nobel Prize in Literature laureate was an honorary doctor at Greifswald University.[63] All in all, there was a continental sociocultural influence[64] and

[58] Ibid, p. 614.

[59] Ibid, p. 615.

[60] Ibid, p. 617f.

[61] Ibid, p. 618.

[62] Ibid, p. 619.

[63] Wolfert 2006, pp. 2–10.

[64] Strömholm 2013, p. 30.

Baltic coast legal influence[65] on the Swedish legal system due to a centuries-long common history. All these interrelations might have left some German impact on the Swedish legal system when following the method proposed by *Kischel* and supported by this thesis.

The following questions lay the groundwork for a comprehensive evaluation:

1. Do the examined legal systems ensure a certain standard of living conditions, including thermal efficiency, for low-income households? If so, how are these assurances implemented?
2. Have low-income tenants encountered legal obstacles concerning the decarbonisation of residences within these legal systems?
3. Have any legal barriers been successfully addressed within these systems? If so, how?
4. Are there non-legal strategies or instruments within these legal systems that might address the landlord-tenant dilemma?

The economic and social context is crucial here, alongside the legal landscape. The private sector could offer financing approaches that, while not directly regulated by the state, play a role in problem-solving.

2.4 Economic Analysis of Law

Given its economic relevance, this research delves into not just the legal but also the economic aspects. A central model in classical economics, pivotal for economic analysis, is the *homo oeconomicus* model. It describes individuals as rational actors striving to maximise personal advantages. They operate with clear preferences and process all market-available information.[66] When considering decisions about energy efficiency enhancements, this rational actor will only opt for such improvements if the anticipated costs of energy use surpass the costs of implementing efficiency measures. The *homo oeconomicus* would carry out all economically reasonable measures to increase the energy efficiency and to realise the associated cost advantages.[67]

When addressing the landlord-tenant dilemma, incorporating a positive analysis can contribute to the assessment of possible solutions for the landlord-tenant dilemma Positive analysis closely aligns with theories regarding human decision-making and actions. Legal economics serves to elucidate the formation of laws, thereby facilitating a more rational legal policy.[68]

[65] Carlson 2012, p. 33.

[66] Grave et al. in: Härtel/ Pielow 2014, p. 68; Rodi 2014, p. 19.

[67] Grave et al. in: Härtel/ Pielow 2014, p. 68.

[68] Rodi 2014, p. 16.

An economic analysis of law aims to integrate economic and legal methods. *Rodi* emphasises that the economic analysis of law extends beyond merely applying economic methodologies to the societal dimensions of law, as the subject matter under review significantly shapes the approach taken in economic analysis. For example, the questions of the fundamental nature of law in a general sense and the objectives of specific legal norms are inherently jurisprudential in nature. Consequently, an assessment of the effectiveness of an ecological tax reform necessitates a probing of the law's purpose in terms of normative impacts, which requires interpretation through legal methodologies.[69]

Comparative law and economics have clearly demonstrated that differences in various legal systems do not necessarily imply inefficiencies. Therefore, different legal traditions may have developed alternative solutions to identical legal issues, which are neutral from an efficiency standpoint.[70] However, the economic approach to law has attracted substantial criticism, particularly for focusing on the "commodification" of legal discourse and neglecting higher values such as ethics, morality, or constitutional principles.[71] Nevertheless, the economic analysis of law appears to be an appropriate method for limiting the scope of this research, particularly regarding the practical applicability of foreign experiences to Germany.

The improvement of thermal efficiency in dwellings entails more than merely providing assets; otherwise, it could be straightforwardly regulated by legal provisions. It aims to "produce" energy efficiency in a market lacking a functioning supply, demand, and corresponding pricing mechanism for this economic good.[72] The economic benefit of energy conservation is often described in literature as an environmental good.[73] In cases where enhanced energy efficiency in a building does not lead to a reduction in heating costs for the user, and where production costs exceed these savings, the investor yields a different return. Conversely, all other consumers gain an economic benefit by receiving more, and therefore cheaper, energy. Like other environmental goods, energy conservation has no exchange or use value. The energy savings generated can be utilised by many without diminishing their utility to others, signifying a universal benefit.[74] As there is no real market for thermal energy efficiency improvement of dwellings, a market price cannot adequately reflect its value. Market players can benefit from energy savings without revealing their preferences and willingness to pay. Energy-saving achieved through such refurbishment is, moreover, non-exclusive because it is impossible to prevent non-paying individuals from using the energy.[75] Low-income tenants are even more sensitive to the

---

[69] Ibid, p. 11.

[70] Caterina in: Smits 2006, p. 163f.

[71] Mattei 2000, p. 53.

[72] Hallof 2013, p. 100.

[73] Ibid; see more: Owen 1999.

[74] Ibid.

[75] Ibid, p. 100f.

split incentives and are more affected by their adverse effects, often lacking the realistic option of relocating to a different flat due to the constrained rental housing market in major cities like Berlin, Munich, Stockholm, etc.

At this point, the positive analysis could contribute to the development of potential solutions for the landlord-tenant dilemma by scrutinising the behaviour of the involved actors and aiding in identifying the most rational decisions for all parties. *Hallof* highlights the critical importance of economic analysis, for example, in the legal comparison of protections against the dismissal of rental contracts. It helps to understand the interests of both landlords and tenants handing behind legal provisions. From a legal point of view, dismissal protection primarily protects the tenant from the inconvenience and costs associated with relocating. At the same time, economics sees two functions of dismissal protection. Firstly, it is an instrument to prevent opportunistic behaviour of the landlord as one of the aspects of tenant protection. Secondly, it serves the interests of a rational landlord in the situation of asymmetric information distribution.[76] In the context of low-income households, which deal with informational barriers among other challenges, modern economic analysis methods assist in comprehending the decision-making mechanisms of each actor involved in dwelling refurbishment. In real markets, decisions are often made with incomplete information. The market failure category of "information deficiency" refers to scenarios where market actors are uninformed to such an extent that the market is significantly impaired and "fails" in its function.[77] Applying this to the researched question, it becomes apparent that discussing a market for energy refurbishment of dwellings is challenging due to an informational imbalance between contracting parties—the landlord and the low-income tenant. However, acknowledging and addressing this issue facilitates the implementation of legal regulations to correct such imbalances. Moreover, economic analysis is crucial if dwelling refurbishment is to be delegated to a third party, not involved in the landlord-tenant dilemma. This approach helps to identify which monetary incentives might motivate independent investors to allocate their assets into the rented dwellings of low-income households. Furthermore, it helps to evaluate the practicability of such an option beyond the scope of this thesis. Based on these insights, recommendations for regulatory policy related to the energy modernisation of rented dwellings can be designed.

---

[76] Hallof 2013, p. 94f.

[77] Rodi 2014, p. 43.

2.5 Limitations

2.5.1 Evaluation Criteria and Their Limitations

The initial step of the evaluation process involves extracting potential solutions, if any, from the context of their respective legal systems. These will be examined in their basic form. The purpose of this extraction is to view the solution solely in terms of its function, evoid of any conceptual context or national doctrine. This approach, proposed by *Zweigert* and *Kötz,*[78] will be adhered to in this thesis. It is imperative, particularly if it is discovered that Sweden addresses the same issue differently, to understand why.

The comparative method helps to construct a system while comparing different legal orders. There are three essential types of relationships crucial for functional comparison:

- the similarity relation between the problems in different societies;
- the functional relation between each specific problem and the corresponding legal institution within the investigated legal system; and finally
- the equivalence relation between the institutions in various legal systems.[79]

Hence, the question posed by functionalist comparison inherently includes a system for further investigation and evaluation.[80] While there is ongoing debate about whether comparative law can serve as a "yardstick" to determine superior law,[81] comparative lawyers agree that evaluating the results of functionalist comparison has its limits. Firstly, the evaluation criteria cannot mirror the criteria for comparability. Secondly, any assessment of functionally equivalent institutions cannot conclude that one is superior to another, as such evaluations are valid only concerning the function scrutinised in the comparative research. Therefore, comparability should be approached with certain restrictions in evaluating results.[82] For this research, it implies that the criteria for evaluation of results need to be developed, taking into account these constraints.

The premise has severe operational limitations related to the similarity of problems across different legal systems. The similarity in functions ceases once a system assigns different social significance to a particular issue. There may be "blind spots" in laws, i.e., some issues are not regulated at all, falling outside the scope of functionalism.[83]

The functional method can also serve as a tool for the unification of law. Identifying similarities among seemingly disparate laws facilitates the drafting of optimal uniform

[78] Zweigert/Kötz 1998, p. 43f.

[79] Michaels in: Reimann/ Zimmermann 2006, p. 373.

[80] Ibid.

[81] Ibid.

[82] Ibid, p. 375.

[83] Brand 2007, pp. 417–418.

legislation, taking into account the nuances of local legal systems. Upon recognising the functional similarities of different legal systems, unification based on these similarities becomes more feasible.[84] European lawmakers seem to favour functional equivalence over unification, as European directives need to be implemented in terms of their results rather than their doctrinal structure. However, this argument is also criticised in the literature. For example, the functional method is sometimes viewed as inadequate at the level of European law.[85]

In terms of critical function, functional analysis assists in overcoming biases against foreign legal systems. Yet, the functional method does not enable one to critique one's own law based on functional equivalence; it is not appropriate to assert whether foreign law is superior. It is possible to identify alternative solutions, but whether these alternatives are better remains uncertain.[86]

There are additional risks in evaluating outcomes related to the epistemological aspects of the functional method. Understanding another legal system, and thus evaluating the outcome, can be challenging due to legal norms being deeply embedded within specific economic, political, moral, and cultural contexts. Moreover, historical perspectives often play a crucial role in comprehending a legal system.[87] Additionally, achieving "cultural fluency" in another legal language when comparing different legal orders presents its own challenges.[88] Furthermore, even if a legal norm or phenomenon is perfectly understood, translating it into one's own language can be problematic, especially if the law is formulated by norms, doctrines, and concepts not precisely mirrored in another legal system.[89]

All these factors limit the evaluation process and, consequently, the outcomes to a certain extent. However, these limitations should be continuously considered throughout the evaluation process to develop a practically applicable outcome and avoid methodological mistakes.

2.5.2 Linguistic and Terminological Limitations

Numerous disciplines e.g., economics, mathematics, physics etc., share a common vocabulary. In contrast, legal terminology varies significantly from country to country, often leading to misunderstandings.[90] While comparing different legal systems, it

[84] Michaels in: Reimann/ Zimmermann 2006, p. 376.

[85] Ibid, p. 377.

[86] Michaels in: Reimann/ Zimmermann 2006, pp. 378–379.

[87] Jansen in: Reimann/ Zimmermann 2006, p. 306.

[88] Ibid.

[89] Ibid, pp. 306f.

[90] De Cruz 2007, p. 220.

becomes necessary to engage with multiple languages and legal translations, i.e., translation from one language into another across different legal systems.[91] Several lexical or terminological challenges arise in legal translation. For this thesis, legal concept issues and legal terms that lack equivalents in another jurisdiction might be relevant.

Firstly, the nature of a legal concept needs clarification. *Cao* defines legal concepts as "abstractions of the generic legal thoughts and rules within a legal system".[92] They help to describe the branches of law and demonstrate which principles, rules and standards apply to them.[93] For instance, German civil law recognises the principle "*Treu und Glaube*" ("the good faith"), codified in §242 BGB. German jurisprudence has developed over the decades a multitude of principles and individual regulations, for instance, the entire law of general notions and conditions up to a codification.[94] The interpretation of this principle in German law may not align with its counterpart in other legal systems.

Cao identifies three dimensions of a legal concept: linguistic, referential, and conceptual. Determining the translatability of a legal concept from one language to another requires assessing its equivalence or similarity in these three dimensions. It is important not to overlook the role of culture-specific criteria in shaping legal concepts.[95] Various methods can be employed for translating legal concepts, including literal translation, formal equivalence, word-to-word translation, functional equivalence, borrowing, and descriptive equivalence.[96]

As an example, the German "*Rechtsstaat*" is often translated as the rule of law. However, the US American interpretation of the rule of law does not fully encapsulate the German understanding of "*Rechtsstaat*". This concept extends beyond the "*rule of law*", which solely assumes that state power should be limited by legal norms that have been introduced as far as possible in advance and, if possible, by the legislature. "*Rechtsstaat*" also encompasses constitutional legality of laws and includes principles of human dignity, freedom, justice, legal certainty, the existence of fundamental rights, legal protections, separation of powers, and the principle of proportionality.[97]

Regarding the translation of legal terms that differ in content across languages, this thesis will not translate terms into English when there is no identical meaning. The original term will be maintained to prevent misunderstandings or incorrect translations, especially when translating from Swedish. However, when linguists translate legal terms into another language, they try to find the best equivalents to avoid losing the legal sense,

---

[91] Doczekalska in: Olsen/ Lorz/ Stein 2009, p. 120.

[92] Cao 2007, p. 54.

[93] Ibid.

[94] Kischel 2015, p. 420.

[95] Cao 2007, p. 55.

[96] Cao 2007, p. 59.

[97] Kischel 2015, p. 191f.

which is deliberately omitted in this paper. The same approach will also apply when a legal term lacks an equivalent in another jurisdiction.

2.5.3 Limitations Related to the Literature Review

Given the dynamic nature of the researched topic and the frequent legislative changes encountered during the course of this thesis, some amendments were adopted so rapidly that the academic literature could not keep pace. For this reason, technical reports, study papers from research institutes or non-governmental organisations, and communications between state and EU bodies were analysed, as these sources were occasionally the only ones available.

The Covid-19 pandemic and its associated restrictions precluded the possibility of a research visit to Sweden. Consequently, the literature review focusing on the Swedish context was confined to resources accessible in Germany. Due to the specificity of legal thinking and the role of the legal science in Sweden described in this chapter, there was often little literature on the researched topic in English and Swedish. Therefore, a big part of this research involved analysing Swedish law directly. The results were then verified through interviews with stakeholders (Swedish tenancy lawyers and *Boverket*—the Swedish National Board of Housing, Building and Planning).

2.5.4 Limitations Related to Replication Potential

When comparing best practices of other countries, one should be critical of their replication potential, even if these are EU members. For example, France has implemented broader measures to address energy poverty among its citizens compared to Germany.[98] However, replicating these practices in the German context presents challenges due to fundamental differences in state organisation between the two countries. France is traditionally a centralised state, exerting significant control over the energy sector, such as through tariff regulation. This control facilitated early implementation of protections for vulnerable customers. In contrast, the issue of energy poverty has been somewhat overlooked in Germany, owing to a robust social protection system, including income support for low-income households' energy consumption. Furthermore, the French energy sector remains highly centralised with substantial state intervention powers, whereas the German energy sector is characterised by decentralisation and division among numerous utility companies.[99]

---

[98] Dubois/ Mayer 2013, p. 1.

[99] Ibid, p. 1f.

References

Artz, Markus, and Ulf Börstinghaus. 2019. 'Artz/Börstinghaus: Das Am 1.1.2019 in Kraft Getretene Mietrechtsanpassungsgesetz'. *Neue Zeitschrift Für Miet- Und Wohnungsrecht*, no. 1: 12–25.

Brand, Oliver. 2007. 'Conceptual Comparisons: Towards a Coherent Methodology of Comparative Legal Studies'. *Brooklyn Journal of International Law* 32 (2): 405–66.

Cao, Deborah. 2007. 'Translating Law'. In *Translating Law*. Multilingual Matters. https://doi.org/10.21832/9781853599552.

Carlson, Laura. 2012. The Fundamentals of Swedish Law: A Guide for Foreign Lawyers and Students. Edition 2. Lund: Studentlitteratur.

De Cruz, Peter. 2007. *Comparative Law in a Changing World*. 3rd ed. London ; Routledge-Cavendish.

Deutscher Bundestag. 2018. 'Gesetzentwurf Der Bundesregierung, Entwurf Eines Gesetzes Zur Ergänzung Der Regelungen Über Die Zulässige Miethöhe Bei Mietbeginn Und Zur Anpassung Der Regelungen Über Die Modernisierung Der Mietsache (Mietechtsanpassungsgesetz – MietAnpG)'. Drucksache 19/4672. http://dip21.bundestag.de/dip21/btd/19/046/1904672.pdf≥.

Dubois, Ute, and Ines Mayer. 2013. 'Energy Poverty in France and Germany. Perception and Policy Approaches, International Association for Energy Economics'. In . Düsseldorf. https://www.semanticscholar.org/paper/Energy-Poverty-In-France-And-Germany%3A-Perceptions-Mayer/7839ca1dc9b0189ac7c439530fbf9dda37acb1a0.

Hallof, Irene J. 2013. *Das Vermieter-Mieter-Dilemma bei der energetischen Gebäudesanierung: eine rechtliche und ökonomische Analyse*. Praxis und Theorie des Bau- und Immobilienrechts 20. Berlin Brüssel: Lexxion Der Juristische Verlag.

Harding, Andrew, and Esin Örücü. 2002. *Comparative Law in the 21st Century*. W.G. Hart Legal Workshop Series. London: Kluwer Academic. http://bvbr.bib-bvb.de:8991/F?func=service%26doc_library=BVB01%26doc_number=009884507%26line_number=0001%26func_code=DB_RECORDS%26service_type=MEDIA.

Härtel, Ines, and Johann-Christian Pielow. 2014. *Effektiv zu mehr Energieeffizienz?: Dokumentation der XV. Jahrestagung des Instituts für Berg- und Energierecht am 19. Mai 2011*. 1st ed. Richard Boorberg Verlag.

Kischel, Uwe. 2015. *Rechtsvergleichung*. München: C.H. Beck.

Mattei, Ugo. 2000. *Basic Principles of Property Law: A Comparative Legal and Economic Introduction*. Westport, Conn: Bloomsbury 3PL.

Olsen, F., R. Lorz, and D. Stein, eds. 2009. *Translation Issues in Language and Law*. 2009th ed. New York: Palgrave Macmillan.

Owen, Gill. 1999. *Public Purpose Or Private Benefit?: The Politics of Energy Conservation*. Manchester University Press.

Pawlowski, Hans-Martin. 1999. *Methodenlehre für Juristen: Theorie der Norm und des Gesetzes ; ein Lehrbuch*. 3., Überarb. und erw. Aufl. C. F. Müller großes Lehrbuch. Heidelberg: Müller.

Reimann, Mathias, and Reinhard Zimmermann. 2006. *The Oxford Handbook of Comparative Law*. Oxford Handbooks. Oxford: Oxford university press.

Rodi, Michael. 2014. *Ökonomische Analyse des Öffentlichen Rechts*. Lehrbuch. Berlin Heidelberg: Springer Gabler.

Saidov, Akmal' Kholmatovich. 2003. *Comparative Law*. Wildy, Simmonds & Hill.

Savigny, Friedrich Carl. 1840. *System Des Heutigen Römischen Rechts*. Vol. I. Berlin.

Schneider, Jens-Peter. 2007. *Verwaltungsrecht in Europa. 1: England und Wales, Spanien, Niederlande*. Schriften zum europäischen und internationalen Recht 14. Göttingen: V-&-R-Unipress.

Smits, J. M., and Edward Elgar Publishing, eds. 2006. *Elgar Encyclopedia of Comparative Law*. Elgar Original Reference. Cheltenham, UK ; Northampton, Mass: Edward Elgar Pub.

Sodan, Helge, and Jan Ziekow. 2014. *Grundkurs Öffentliches Recht: Staats- und Verwaltungsrecht*. 6., Wesentlich überarbeitete Auflage. München: C. H. Beck.

Stromholm, Stig. 2013. *An Introduction to Swedish Law: Volume 1*. Softcover reprint of the original 1st ed. 1981 Edition. Dordrecht: Springer.

Tolochko, Alexander. 2014. 'Modern Typological Model of Criminal Process of Ukraine: A Comparative Legal Analysis'. *Austrian Journal of Humanities and Social Sciences*, 271–76.

Wolfert, Raimund. 2006. 'A History of German – Scandinavian Relations'. https://www.bbsr.bund.de/BBSR/DE/veroeffentlichungen/izr/2009/8_9/Geschichte_engl.pdf?__blob=publicationFile&v=1.

Zweigert, Konrad, and Hein Kötz. 1996. *Einführung in die Rechtsvergleichung auf dem Gebiete des Privatrechts*. Mohr.

Piek, Sarah. 2013. 'Die Kritik an der funktionalen Rechtsvergleichung'. Zeitschrift für Europäische Privatrecht, 60–87.

Current State of Research in Defined Area from Energy Efficiency Point of View

3

3.1 Background of Low-income Households

3.1.1 Poverty and Energy Poverty Definitions

Before discussing the landlord-tenant dilemma within the context of low-income households, it is important to have a clear understanding of this term. In Germany, there is no legal definition of "low-income households", necessitating consideration of this phenomenon's non-legal context. This will later contribute to the legal comparison of the regulatory framework in the energy modernisation of housing occupied by low-income tenants. Moreover, understanding the differences within a low-income group is a necessary precondition for developing and implementing energy efficiency policies. Targeting low-income households, due to their social constraints, could optimise the benefits of energy efficiency programmes.[1] Improved energy efficiency in low-income households is not only an irreversible component of reducing energy poverty but also the contribution to CO_2 reduction and climate change policies.[2]

A "low-income" household does not necessarily equate to a "poor" one, particularly when comparing households in different countries. For instance, an unemployed person in Germany receiving state aid may have more financial resources than an employed person in a developing country. However, this does not imply that the unemployed individual is poor, nor does it mean that the employed person with a steady income belongs to

[1] Podgornik et al. 2016, p. 26.

[2] Boardman 2010, p. 125.

J. Karras, *Legal Barriers to the Energy Modernisation of Dwellings Occupied by Low-income Tenants and Opportunities to Overcome these Barriers*, Schriftenreihe des Instituts für Klimaschutz, Energie und Mobilität,
https://doi.org/10.1007/978-3-658-44193-7_3

a low-income group. The definition of poverty serves as an indicator for determining the low-income threshold in a country. Therefore, defining poverty is essential. Yet, researchers do not agree on how to measure poverty, as the extent of poverty depends on various factors, including the poverty definition and the methodology applied.[3]

Researchers identify two types of poverty: relative and absolute. Absolute poverty affects individuals who cannot meet their basic essential needs, thereby threatening their physical existence.[4] This form of poverty is rare in modern societies, occurring only in extreme cases such as the risk of starvation or freezing. *Hradil* criticises definitions that focus solely on absolute poverty, arguing that appropriate concepts of poverty should relate to the socio-cultural subsistence level of people, as determined by the historical, political, and materially shaped evaluations of a society.[5]

Relative poverty refers to a social condition where individuals, families, or groups have so few resources that they are excluded from the minimum acceptable way of life in their country.[6] This definition is similar to the one used in the report of the European Council and the Council of the European Union in 2003.[7] According to this definition, people are affected by poverty "if their income and resources are so inadequate as to preclude them from having a standard of living considered acceptable in the society in which they live. Because of their poverty, they may experience multiple disadvantages through unemployment, low income, poor housing, inadequate health care and barriers to lifelong learning, culture, sport and recreation. They are often excluded and marginalised from participating in activities (economic, social and cultural) that are the norm for other people, and their access to fundamental rights may be restricted."[8]

The method of measuring a minimum acceptable lifestyle is subject to debate. Income distribution directly influences the level of poverty within a state. Firstly, the relative income poverty line is established to measure poverty in a specific state. The relative income poverty line is calculated in relation to the average or median household income of each state under study, while also considering household size and composition.[9] At the same time, the European Commission sets up the at-risk-of-poverty threshold at 60% of the national median equivalised disposable income. For cross-country comparison, the purchasing power standards are used to account for cost of living variations across countries.[10]

[3] Köhler 2016, p. 1.

[4] Blaser in: Dabrowski/ Wolf 2018, p. 41f.; Karras/ Novikova 2021, p. 1.

[5] Hradil 2005, p. 244; Karras/ Novikova 2021, p. 1.

[6] Blaser in: Dabrowski/ Wolf 2018, p. 43; Karras/ Novikova 2021, p. 2;

[7] COM (2003) 773 final.

[8] COM (2003) 773 final; Dvorak in: Odekon 2015, p. 528.

[9] Dvorak in: Odekon 2015, p. 528.

[10] 'Archive:Income Poverty Statistics'.2018. 2018; Karras/ Novikova 2021, p. 2.

The decision to define 60-limit as an EU poverty indicator is instead a political one. Nevertheless, some scholars demand a poverty risk rate instead of a poverty rate.[11] The mere consideration of poverty fails to provide nuanced insights into the Commission's measures and deficiencies in other aspects of life. The focus is not strictly on poverty per se but on the heightened risk of impoverishment. This context often leads to discussions around income poverty, which neglects the consideration of individual assets. Additionally, other life circumstances that might influence a legitimate assessment of potential poverty are not accounted for within the scope of income poverty. However, this method does not evaluate risk but rather an actual relative insufficiency of income. This deficiency has a decisive influence on opportunities to participate in almost all areas of social life.[12] The relevance of this approach to poverty, particularly in modern Western European, market-based societies, is underscored by the fact that almost all basic needs are monetarily attainable.[13] This understanding of poverty should be kept in mind when discussing low-income households.

Another issue is the partial invisibility of poverty. National statistics mainly capture those with a registered office address, leaving unregistered individuals and undocumented migrants unaccounted for in the data.[14] However, this demographic is not the focus of this research, as they typically do not engage in renting dwellings. Therefore, this segment of the impoverished population remains outside the scope of this thesis.

Other poverty indicators are often country-related. Climate may also serve as an indicator of energy poverty. Interestingly, energy poverty tends to be a more significant issue in warmer countries than in colder ones. The colder climate made countries develop more energy-efficient dwellings and improve building standards and technologies progressively.[15] In contrast, for warmer countries, such advancements have only become necessary in recent years.

Another indicator could be social assistance paid from public means. For example, in Germany, a segment of the population can only attain the socio-cultural subsistence minimum, a constitutional right, through basic income support for job seekers in accordance with Book II of the German Social Code (SGB II) and social aid under Book XII of the German Social Code (SGB XII), or through benefits provided under the Asylum Seekers Benefits Act. They are recipients of minimal social security benefits from the state means. In August 2022, around 5.4 million people in Germany received benefits under the basic security scheme for jobseekers (ALG II).[16] It can be assumed that absolute

---

[11] Blaser in: Dabrowski/ Wolf 2018, p. 44.

[12] Ibid.

[13] Hradil 2005, p. 243.

[14] European Anti-Poverty Network. 2019. 'Poverty Report Sweden 2019'. https://www.eapn.eu/wp-content/uploads/2020/04/EAPN-PW2019-Sweden-EN-EAPN-4306.pdf ., p. 1.

[15] 'Energy – European Commission'.2023. 19 December 2023. https://energy.ec.europa.eu/index_en.

[16] Bundestag, Drucksache 20/3873, p. 1.

poverty is effectively mitigated through social assistance for subsistence. Moreover, social assistance aims to ensure a dignified life. Initially, the welfare system was perceived as a component of police and regulatory law, and an individual's legal entitlement to social assistance was not recognised. However, shortly after the Basic Law was enacted, the Federal Administrative Court deduced a constitutional mandate to establish the minimum conditions for a dignified existence, particularly from the principle of human dignity in Article 1 (1) of the Basic Law (Grundgesetz—GG).[17] The principle of the welfare state under Article 20 (2) of the Basic Law also requires that the minimum subsistence level be secured through state social assistance benefits. The number of individuals impacted by absolute poverty who are either unwilling or unable to claim social assistance is fortunately very small.

In Sweden, the situation appears more favourable at first glance, especially when considering absolute figures. The proportion of people at risk of poverty or social exclusion is lower than the EU average.[18] According to Statistics Sweden, only 1.6% of the Swedish population lived in severe material poverty in 2018, equating to just over 160,000 individuals.[19] The largest group experiencing material poverty in Sweden comprises unemployed persons. Furthermore, people with a migration background face a higher risk of severe material poverty compared to domestic-born individuals with both parents born in Sweden, with rates from 4% to just under 1% in 2018.[20] Additionally, 16.4% of Swedish people lived in relative poverty in 2018 i.e., with a low economic standard, representing nearly 1.4 million people. Over the past decade, this percentage has increased from just over 13% to over 16%, due to various factors including rising income disparities.[21]

Both Germany and Sweden, with their developed economies and high standards of living, ensure that even individuals receiving social benefits can satisfy more needs than employed persons in some EU member states. For example, the standard rate for a single person (standard level 1) of ALG II in Germany was EUR 449 in 2022.[22] Additionally, the state covers the actual costs for accommodation and heating, as long as they are

---

[17] BVerwGE 1, 159.

[18] COM (2019) 150 final, p. 10.

[19] European Anti-Poverty Network. 2019. 'Poverty Report Sweden 2019'. https://www.eapn.eu/wp-content/uploads/2020/04/EAPN-PW2019-Sweden-EN-EAPN-4306.pdf.

[20] Ibid.

[21] Ibid.

[22] 'Erhöhung Der ALG II Regelsätze Ab 2022 – Jobcenter Heidenheim'. n. d. Accessed 20 December 2023. https://www.jobcenter-heidenheim.de/aktuelles/erhohung-der-alg-ii-regelsatze-ab-2022/.; Since 01.01.2023, ALG II social benefit has been replaced with the *Bürgergeld* (citizens' income). This is a major social reform of the Federal government that shall make the state assistance closer to the citizen, less bureaucratic and more targeted, see more: 'Bürgergeld ersetzt ALG II | Bundesregierung'. 2023. Die Bundesregierung informiert | Startseite. 1 January 2023. https://www.bundesregierung.de/breg-de/schwerpunkte/entlastung-fuer-deutschland/buergergeld-2125010.

deemed reasonable, based on local housing market rents. Social assistance in Germany also includes medical insurance coverage.[23] In contrast, the average monthly gross salary in Bulgaria in 2022 was €852.[24] Nonetheless, people receiving social benefits cannot afford an average standard of living in their countries.

Low-income employees, students, trainees and pensioners are also often at risk of poverty, even if they do not receive any social benefits, since their total income often is below or at the level of 60% of the median needs-weighted net income of the population in private households in Germany. In 2018, individuals in Germany earning less than €13,628 annually were considered to be living at risk of poverty.[25] In 2019, the at-risk-of-poverty rate in Germany was 15.9%. In eastern Germany, this rate was around 17.9%.[26] A similar rate existed in Sweden in 2018, with 18%, or approximately 1.8 million people, living there at risk of poverty or social exclusion. This figure is also risen by 1.3% during the last decade.[27] It is expected that this rate will increase in both countries due to the economic crisis caused by the Covid-19 pandemic.

The concept of capability warrants mention in this context, as both Sweden and Germany possess high living standards, which positively influence the fulfilment of needs among low-income individuals compared to their counterparts in developing countries. This concept, developed in the 1980s, defined what constitutes a "good life". According to this concept, the individuals concerned should reasonably define the possibility of living a good life. Consequently, their life context should also be taken into account. The concept focuses not solely on the availability of resources or subjective well-being. Therefore, low-income households should not be considered in isolation but rather in comparison with other social groups within the same country.[28] Following this concept, poverty can be understood as "deprivation to live a good life".[29] It is important to note that vulnerability does not equate to deprivation in this context.[30]

---

[23] 'Regelsätze steigen zum 1. January 2021'. 2021. Die Bundesregierung informiert | Startseite. 1 January 2021. https://www.bundesregierung.de/breg-de/service/archiv/regelsaetze-steigen-1775798.;

[24] 'Average Monthly Salary in European Union 2022 | Reinis Fischer'. n. d. Accessed 20 December 2023. https://www.reinisfischer.com/average-monthly-salary-european-union-2022 .

[25] 'Themenseite: Armut in Deutschland'. n.d. Statista. Accessed 20 December 2023. https://de.statista.com/themen/120/armut-in-deutschland/.

[26] 'Armutsgefährdungsquote in Deutschland bis 2022'. n. d. Statista. Accessed 20 December 2023. https://de.statista.com/statistik/daten/studie/72188/umfrage/entwicklung-der-armutsgefaehrdungsquote-in-deutschland/ .

[27] European Anti-Poverty Network. 2019. 'Poverty Report Sweden 2019'. https://www.eapn.eu/wp-content/uploads/2020/04/EAPN-PW2019-Sweden-EN-EAPN-4306.pdf.

[28] Bartiaux/ Vandeschrick/ Moezzi/ Frogneux 2018, p. 1221 f.

[29] Ibid, p. 1222.

[30] Ibid.

In summary, as mentioned earlier, low-income households in the German and Swedish contexts are not only those at risk of poverty and social exclusion but also those living above the poverty line who still cannot afford to satisfy an average amount of needs, thereby struggling to live a good life for various reasons. For instance, elevated rental costs in major cities like Munich, Frankfurt, or Stockholm impose a burden on numerous employees, even those earning above the poverty risk threshold.

Low-income individuals often face difficulties in paying their energy bills, cannot afford to rent energy-efficient dwellings, and do not own their properties. Increasing energy prices become a heavy burden for energy bills of this society groups. All these factors influence their risk of social exclusion and poverty. They are also indicators of energy poverty. However, these terms are related to each other but are not identical.[31]

In 2022, private households in Germany spent, on average, 8,04 cents per kilowatt-hour on natural gas from January to June. They were charged an average 33.50 cents per kilowatt hour of electricity. Compared with the second half of 2021, gas prices increased by 17.7% and electricity prices by 1.9%.[32] These circumstances created an additional burden for low-income people.

Unfortunately, not everyone can afford to live in energy-efficient housing, even in developed countries. An essential but often overlooked residential energy sector is that of low-income households. Official statistics indicate that 16.8% of the EU's total population was at risk of poverty or social exclusion in 2018.[33] These numbers are expected to increase in light of the migrant crisis and Covid-19 pandemic effects in the subsequent years. This vulnerable group typically resides in older buildings with lower energy performance, leading to increased energy consumption and higher energy bills. They are unable to purchase affordable warmth,[34] resulting in a vicious circle. Therefore, this group is a suitable group to target for energy-saving initiatives.[35] Numerous factors discussed in the literature contribute to the burden of energy costs in low-income households: energy prices, the low thermal energy efficiency of buildings they live in, socioeconomic demographics as well as behavioural patterns of residents.[36]

Low-income households often struggle with energy poverty, i.e. the situation where they cannot adequately heat or provide other required energy services in their homes at affordable costs.[37] Nearly 34 million people in the EU could not afford to heat their

---

[31] Karras/ Novikova 2021, p. 2.

[32] 'Household Gas Prices up 17.7% in 1st Half of 2022'. n. d. Federal Statistical Office. Accessed 20 December 2023. https://www.destatis.de/EN/Press/2022/10/PE22_460_61243.html.

[33] 'Archive:Income Poverty Statistics'. 2018. 2018. https://ec.europa.eu/eurostat/statistics-explained/index.php?title=Archive:Income_poverty_statistics.

[34] Bouzarovski/ Tirado Herrero 2017, p. 73.

[35] Chen et al. 2017, p. 61.

[36] Ibid.

[37] Pye/ Dobbins et al. 2015, p. V.

homes adequately in 2018.[38] This problem has been described as fuel (energy) poverty in literature.[39] Both terms are used synonymously.

Energy poverty is a multi-dimensional concept. It would be incorrect to assert that it results solely from a combination of low income and energy-inefficient homes, although these are two primary indicators. The EU Energy Poverty Observatory (EPOV), a consortium of 13 organisations, including universities, think tanks, and the business sector, has identified primary and secondary general indicators of energy poverty to assist member states in addressing the issue. Primary indicators include arrears on utility bills, low absolute energy expenditure, a high share of energy expenditure in income and inability to keep the home adequately warm. Secondary indicators encompass fuel oil, coal and biomass prices, household electricity, district heating and household gas prices, dwelling comfortably cool during summer times and warm during winter times, a number of rooms, persons, owners, location of dwellings, poverty risk, energy label of dwellings, energy expenses, condition of dwellings, kind of equipment for warmth as well as energy expenses.[40] However, the quantity of energy used in the dwellings and its cost are central to the definition of energy efficiency in this context. Thus, three core components determine fuel poverty: income, fuel prices, and the energy efficiency of the dwellings. Additional contributing factors include a lack of financial assets and tenant status, which limit a household's ability to improve its property.[41]

There is no uniform definition of energy poverty. According to the Commission Recommendation on Energy Poverty (COM (2020) 9600 final), energy poverty is "a situation in which households are unable to access essential energy services", recital 1. However, the Recommendation has no binding force on the member states. Moreover, different terms are used to describe the same economic-sociological problem. North Europe uses the term "fuel poverty" more often, whereas less developed countries tend to use the term "energy poverty" as it addresses a broader spectrum of problems, such as infrastructure, electrification etc.[42] The EU has also adopted the term "energy poverty"[43] as well, as the members of the EU are not only developed countries of North Europe but also still developing countries dealing with a brighter spectrum of issues related to this phenomenon.

Broadman was a pioneer in researching poverty related to affordable warmth. She introduced the term and definition of fuel poverty in 1991,[44] defining fuel-poor families

[38] COM (2020) 950 final.

[39] Boardman 2010, p. 21.

[40] Thema/Florin 2020, p. 6f.

[41] Boardman 2010, p. 21.

[42] Bartiaux/ Vandeschrick/ Moezzi/ Frogneux, p. 1220.

[43] COM (2019) 640 final; COM (2020) 950 final; Commission Recommendation (EU) 2020/1563.

[44] Boardman 1991.

as those whose fuel expenditure exceeds 10% of their income.[45] The UK and Ireland began to research and address the problems of cold and energy-inefficient housing in the early 1990s. Recently, these issues have gained prominence in post-socialist countries of Eastern and Central Europe, as well as in Western and Southern Europe, Australia, New Zealand, and even the United States.[46]

In its Recommendation (EU) 2020/1563 on energy poverty, the Commission defines it as "a situation in which households are unable to access essential energy services".[47] However, this definition is too general to reflect the main problems related to this phenomenon. Hence, a more precise definition is necessary. Therefore, the Commission recommends that member states develop a document providing indicators of energy poverty and defining what constitutes a significant number of energy-poor households, as per paragraph 2 of the Recommendation (EU) 2020/1563.

In the context of the correlation between low income and energy poverty, the Belgian experience, as an EU member and a federal state with a multi-level administrative system, should be considered. In Belgium, a high proportion of dwellings are rented; about 30% of households rented their housing in 2020.[48] In the same year, 22.1% of Belgians were at risk of poverty and social exclusion.[49] These data demonstrate a similar outset in Belgium and Germany, making the Belgian experience particularly relevant.

Belgium began researching energy poverty earlier than many other Western European countries, starting in the early 2000s. To define energy poverty, the Belgian Federal Institute of Statistics, in collaboration with a commercial company, conducted face-to-face interviews with 7163 respondents aged 18 to 80 between 2008 and 2010. In each sampled household, one respondent was interviewed. The sample excluded elderly persons aged over 80 living alone or with other persons all aged over 80, who represented 5.5% of all households in Belgium as of 1 January 2008. Bartiaux et al. analysed these samples and categorised them into three types based on access to affordable warmth:

- energy-poor households
- energy-vulnerable households and
- energy-rich households.[50]

---

[45] Boardman 2010, p. 21f.

[46] Bouzarovski/ Petrova 2015, p. 32.

[47] Commission Recommendation (EU) 2020/1563.

[48] 'Housing in Europe – House or Flat – Owning or Renting'. n. d. Housing in Europe. Accessed 20 December 2023. https://ec.europa.eu/eurostat/cache/digpub/housing/bloc1a.html.

[49] Statbel. 'More than 2 Million Belgians at Risk of Poverty or Social Exclusion'. P – Armoederisico's 2019. Accessed 20 December 2023. https://statbel.fgov.be/en/themes/households/poverty-and-living-conditions/risk-poverty-or-social-exclusion.

[50] Bartiaux/ Vandeschrick/ Moezzi/ Frogneux 2018, p. 1224.

Energy-poor households struggled either to keep their homes adequately warm or to pay their utility bills in the last 12 months, or both. Energy-vulnerable households faced difficulties in keeping their homes warm, primarily due to poor insulation. Some had fallen behind on their utility bills in the past year. Energy-rich households had no issues affording adequate warmth, even if they had arrears in utility bills.[51]

Given Belgium's social security system, some households receive publicly funded aid when other social protections (such as family allowances, unemployment benefits, or benefits for permanent inability to work due to health conditions) are deemed insufficient. Bartiaux et al. identified two additional categories within this typology with the inclusion of this publicly funded aid:

- Aided poorest households and
- other households self-perceived as poor.[52]

The first group found that making ends meet was very difficult or challenging, and they received publicly funded aid or advice from publicly funded municipal centres. The second group also found that it was difficult or somewhat difficult to make ends meet; however, they did not receive the financial aid of last resort.[53] Energy-poor households, the aided poorest households, and others self-perceived as poor have a common feature—they are especially vulnerable to energy and cannot afford an affordable warmth by themselves without sacrificing other needs.

The results of the study conducted by Bartiaux et al. can apply to Germany and Sweden. Belgium has a relatively high share of rented housing—the share of tenants was 27.3% among the total population of Belgium in 2017.[54] It is also a welfare state with a broad spectrum of social aids for those at risk of absolute poverty.

The European Union addressed energy poverty in the European Green Deal.[55] The Commission emphasised that "the risk of energy poverty must be addressed for households that cannot afford key energy services to ensure a basic standard of living".[56] The European Commission also used this definition in its Recommendation on energy poverty of 14 October 2020.[57] The Commission aims to produce guidance to assist its member states in addressing the issue of energy poverty. Thus, for the Commission, two factors are essential in the context of energy poverty—the affordability of essential

---

[51] Ibid.

[52] Ibid, p. 1224f.

[53] Ibid.

[54] 'Belgium: Share of Tenants 2008–2017'. Statista. Accessed 20 December 2023. https://www.statista.com/statistics/544645/tenants-among-population-belgium/.

[55] COM (2019) 640 final.

[56] Ibid.

[57] COM(2020) 9600 final; Karras/ Novikova 2021, p. 2.

energy services and a basic standard of living. It also recognises the connection between the renovation of private buildings and the reduction of energy poverty. The recommendations further identify energy poverty as one of the central elements of the European Green Deal and numerous other pieces of EU legislation and actions, e.g., the Energy Efficiency Directive, the Generation EU Recovery Package and the Cohesion Policy.[58]

The "Fit for 55" legislative package, one of the EGD's following packages, also addresses energy poverty. According to the European Commission, up to 34 million people in the EU have been affected by energy poverty in 2021.[59] This number is greater than the population of some member states, indicating that this is not a phenomenon typical only to some member states but rather a phenomenon relevant to the entire EU and all the member states. Even economically strong countries such as Germany and Sweden are affected by energy poverty. The context of this phenomenon will be analysed in detail below.

However, member states are no longer required to address energy poverty, and Germany has not yet directly addressed the problem of energy poverty. The Third Energy Package obligated the member states to recognise and define social groups and vulnerable consumers in the retail energy markets with respect to meeting their household energy needs.[60] In the German National Energy and Climate Plan (NECP), energy poverty has been discussed. According to the NECP, the Federal Government pursues a comprehensive approach to poverty reduction in social law. Nevertheless, energy poverty is neither defined at the federal level nor at the local level. In Germany, persons in need of assistance have a constitutional right that guarantees the minimum subsistence level in human dignity according to Art. 1 (1) GG in conjunction with the principle of the welfare state under Art. 20 GG. This includes those financial requirements that are indispensable for physical existence and a minimum level of participation in social, cultural and political life. The Federal Constitutional Court has stated that statutory entitlements of unlimited duration secure the guarantee of the minimum subsistence level in human dignity. These are regulated by SGB II and SGB XII (second and twelfth books of the Social Code) and are known as minimum security schemes. These entitlements to subsistence benefits are designed to cover the entire subsistence needs. As subjective rights, they are enforceable and subject to judicial review. The legislature specifies the scope of these entitlements considering the types of need and the resources required therefor. Energy needs are part of the minimum subsistence level and are covered in the aforementioned minimum security systems.[61]

Reasonable costs for heating energy (including hot water production in the case of central heating energy supply) are fully covered under §22 (1) SGB II, §35 (4) SGB

---

[58] Karras/ Novikova 2021, p. 2.

[59] COM (2021) 550 final.

[60] Pye/ Dobbins 2015, p. 26.

[61] Bundesministerium für Wirtschaft und Energie. 2020, p. 67; Karras/ Novikova 2021, p. 2.

XII.[62] Since the citizens' benefit social reform, a one-year transition period applies, during which needs for accommodation are generally recognised in the amount of the actual expenses, §22 (1) S. 2 SGB II. However, if the expenses for accommodation and heating increase after an unnecessary move, only the previous need is recognised, §22 (1) s. 6 SGB II. This provision poses an additional burden for those households willing to move into a better dwelling.

The minimum security systems cover household energy without decentralised hot water production as a lump sum within the framework of the legal needs (cf. §§20 (1) SGB II, 27a (2) SGB XII). The amount of the standard needs is derived from the concrete consumption expenses of low-income households that are recognised as relevant to everyday needs and do not exclusively receive subsistence benefits according to SGB II and SGB XII. Additional needs are also recognised for the decentralised production of hot water (§§ 21 (7) SGB II, 30 (7) SGB XII).[63]

According to German social law, the specific group of persons entitled to assistance is determined by comparing the legislated subsistence level with the income and asset situation of the individual seeking assistance. The NECP notes that an abstract definition of groups affected by "poverty" or "energy poverty" would not suffice for the constitutional requirements of the minimum subsistence level in Germany. The poverty risk rate and primary indicators proposed by the EU Energy Poverty Observatory do not provide information on individual needs, as per the NECP.[64]

As mentioned above, poverty and energy poverty are not identical phenomena, a distinction that is unfortunately overlooked in the NECP 2020. Some organisations have attempted to define energy poverty. This position, however, automatically excludes individuals who struggle to pay energy bills but do not receive subsistence benefits.[65] The Consumer Office of Rhineland-Palatinate (*Verbraucherzentrale Rheinland-Pfalz*) has made a notable attempt, defining energy poverty based on its effects on consumers: a household is considered energy poor if it can only afford energy bills for a standard of living usual in Germany with considerable effort or if it must forgo other basic needs, such as accruing rental debt or reducing spending on nutrition, medicines, and education.[66] Similarly, the Consumer Organisation of North Rhine-Westphalia defines energy poverty as when a household spends an above-average portion of its income on heat and electricity or is unable to pay its energy bills.[67] While these attempts are not entirely

---

[62] Bundesministerium für Wirtschaft und Energie. 2020, p. 67.

[63] Ibid.

[64] Bundesministerium für Wirtschaft und Energie. 2020, p. 67.

[65] Karras/ Novikova 2021, p. 2.

[66] Kalheber, in: Großmann/Schaffrin/Smigiel 2017, p. 209; Karras/ Novikova 2021, p. 2.

[67] Verbraucherzentrale NRW 2015, p. 23; Karras/ Novikova 2021, p. 2.

coherent, a common feature in both definitions is the higher proportion of income spent on energy by those affected by energy poverty compared to average households.[68]

The debate on energy poverty in Germany has been ongoing since the Hans Böckler Foundation began examining the social consequences of rising energy costs for low-income households in 2006. This project was initiated due to a Forsa survey in September 2006,[69] which revealed that almost two-thirds of surveyed households were concerned about rising heating costs in the upcoming winter, with 76% of households earning less than €1000 per month being deeply concerned.[70]

This confirms that there is no uniform definition of energy poverty applicable to all countries. Some households maintain acceptable living standards and low fuel costs despite low income, perhaps due to below-average living costs in their community. Others experience temporary energy poverty, such as during job searches after completing studies. Although the EU has required member states to address energy poverty for several years, Germany has yet to develop its own definition.

In this thesis, a definition of energy poverty in Germany has been developed, drawing upon the discussions and findings previously mentioned. This definition is designed to facilitate the reader's understanding: *Energy poverty is characterised by the inability to either keep the house adequately warm or to pay utility bills in the last 12 months, or both; or reliance on social (publicly funded) assistance that covers energy bills to a reasonable extent, coupled with the necessity to conserve energy to avoid exceeding the amount covered by public funds; or challenges in maintaining adequate warmth in the household without sacrificing other needs, alongside difficulties in making ends meet, even when the individual does not receive (publicly funded) social aid.*

This comprehensive definition, formulated for the purpose of this thesis, aims to address the issues related to affordable house warmth for various vulnerable groups, including those who are not beneficiaries of social aid or who regularly pay their energy bills at the expense of other needs. It seeks to minimise the risk of these groups becoming marginalised in the energy transition, thereby preventing a social divide by ensuring they are not excluded from energy modernisation assistance for their housing due to not being categorised as "poor enough".

3.1.2 Correlation Between Low-income and Energy Efficiency Modernisation

In the past, EU energy policy primarily focused on energy market liberalisation, single market integration, and innovation in energy supply technologies. However, the

[68] Karras/ Novikova 2021, p. 2.

[69] Schreiner 2015, p. 10.

[70] Forsa 2006, p. 14–16.

direction of EU energy policy has shifted towards energy efficiency, now an increasingly important area.[71] Energy efficiency forms a key element of the EU's energy strategy, significantly impacting climate change mitigation. In this context, buildings have been identified as a sector with substantial potential for energy savings.[72] This recognition has been incorporated into the EU Energy Efficiency Plan 2011, and numerous documents in recent years have underscored the need for improvement in energy efficiency. The Energy Union established in Regulation (EU) 2018/1999,[73] has embedded the "energy efficiency first" principle into legislation. Data from 2020 suggests that while the COVID-19 pandemic significantly impacted the achievement of the 2020 energy efficiency targets, it did not lead to a structural reduction in energy consumption, indicating potential rebound effects post-economic recovery.[74] Consequently, modernising and enhancing the energy performance of buildings positively affects the environment due to energy savings and greenhouse gas emission reductions. Moreover, improving building energy performance offers various economic benefits for consumers, public budgets, and energy systems, including enhanced grid stability, reduced network losses, and lower system upgrade costs.[75]

At the same time factors such as poor insulation, outdated window frames, and single glazing contribute to uncomfortable thermal conditions, potentially leading to numerous health issues. Enhancing the envelope components of buildings,[76] including walls, roofs, and glazing, can significantly improve energy efficiency. Given the rising energy prices, vulnerable groups are expected to face increased hardships due to energy-inefficient housing, as escalating energy bills present a substantial burden, and state assistance covers only reasonable costs. According to 2021 statistics, housing costs were considered an excessive burden for 10.7% of the German population.[77]

Energy modernisation emerges as one of the most cost-effective solutions for these vulnerable groups, particularly given that approximately three-quarters of the housing stock in 2011 was constructed prior to the implementation of current thermal insulation

[71] Rosenow/ Kern in: Leal-Arcas/ Wouters 2017, p. 501.

[72] Podgornik et al. 2016, p. 25

[73] Regulation (EU) 2018/1999 of the European Parliament and of the Council of 11 December 2018 on the Governance of the Energy Union and Climate Action.

[74] COM (2020) 950 final.

[75] DG IPOL 2016, p. 22.

[76] "building envelope" means the integrated elements of a building which separate its interior from the outdoor environment (Art. 2(1) point 7 of the Directive 2010/31/EU).

[77] Destatis: Housing costs: 10,7% of the population regarded as overburdened in 2021, 'Housing Costs: 10.7% of the Population Regarded as Overburdened in 2021'. n. d. Federal Statistical Office. Accessed 20 December 2023. https://www.destatis.de/EN/Press/2022/08/PE22_N054_61.html .

policies.[78] In the last decade, the proportion of new buildings constructed in accordance with these policies has increased. However, the issue with old buildings remains unresolved. Moreover, the requirements for insulation were significantly tightened only towards the end of the 2000s through amendments to the Energy Saving Ordinance (Energieeinsparverordnung, EnEV), partly motivated by the EU's Building Efficiency Directive of 2003. Prior to 2007, existing buildings were exempted from subsequent insulation requirements for economic reasons.[79] Therefore, improving insulation in these older buildings is a priority, particularly for low-income groups.[80] Additionally, housing retrofits may also impact social inclusion by alleviating the isolation felt by individuals embarrassed by their living conditions. Enhanced living conditions can lead to reduced frequency of moving, thereby improving school and work attendance, as well as overall household comfort and safety.[81]

In the context of public funding for the energy modernisation of existing housing stock, it is crucial that the policies governing the distribution of these resources are carefully considered, particularly in terms of social sustainability. Within the energy efficiency renovation context, two dimensions of social sustainability are relevant: the preservation of social capital and local communities, and the creation of mixed communities. The first dimension addresses the concern that renovations should not force households, especially low-income or socially vulnerable ones, to relocate. After the renovation, there exists a risk that these households may be displaced, replaced by economically stronger ones. This displacement could result in a decline in social trust, particularly if landlords utilise public funds for energy modernisation. While financially beneficial in the short term, such actions could prove counterproductive for societal cohesion in the long run.[82] The second dimension focuses on fostering mixed communities, characterised by low segregation between various income and ethnic groups. This approach could enable poorer social groups to remain in, or extend their stay within, these areas. However, this dimension extends beyond mere financial considerations, encompassing sociological and urban planning aspects, and thus lies outside the scope of this research.

As mentioned earlier, low-income households often have minimal personal resources and limited access to commercial loans or other forms of capital. Consequently, paragraph 9 of Commission Recommendation (EU) 2020/1563 advises member states to prioritise low-income households when allocating public funds, particularly grants. Nevertheless, challenges may arise even when low-income tenants receive grants for renovation. A notable issue is the potential lack of interest from landlords in participating

---

[78] Koch 2011, p. 644.

[79] Ibid.

[80] Kolokotsa/Santamouris 2015, p. 326.

[81] DG IPOL 2016, p. 24.

[82] Annadotter et.al. 2014, p. 5.

in such renovations. The process of renovating rented dwellings, particularly regarding energy modernisation, frequently involves a landlord-tenant dilemma, which will be explored in the subsequent sub-chapter.

3.2 The Theoretical Background of the Landlord-tenant Dilemma

Tenure status and a dwelling type play a crucial role in tackling the challenges of energy poverty.[83] As buildings account for approximately 40% of Europe's total energy consumption,[84] it is essential to improve the sustainable renovation of buildings. The escalation in energy prices has not only heightened the economic significance of energy consumption but has also led to a shift in mentality towards energy efficiency. Increasingly, more individuals are becoming aware of the importance of energy efficiency. Despite existing efficiency requirements under European law, a significant challenge remains: motivating key decision-makers, such as building owners, property investors, and facility managers, to improve energy performance. This challenge is intricately linked to several questions: 1)What incentives can motivate building owners or facility managers to renovate buildings for better energy performance; 2)How can sustainable retrofitting and renovation be effectively introduced into the market; and 3) What economic incentives can further promote sustainable retrofitting and renovation.[85] The answers to these questions are influenced by factors such as ownership, occupancy, building types, climate, and national context.[86]

Before analysing the structure of the landlord-tenant dilemma, it is important to briefly explore the housing market and the dwelling as an economic good. Economically, the housing market facilitates the exchange of rights to use dwellings and the remuneration for these rights.[87] The unique characteristics of a dwelling as an economic good necessitate specific state housing policies, especially regarding rented dwellings.[88] Primarily, housing, being a basic human necessity, implies that individuals are inherently dependent on having a home.[89] Buildings, and consequently dwellings, are among the longest-lived components of the economy,[90] typically having a lifespan of several decades. They can be occupied by multiple households over time and may be repeatedly

[83] Pye/Dobbins 2015, p. 12.

[84] Recital 3, Directive 2010/31/EU.

[85] Ástmarsson et al. 2013, p. 355.

[86] Ástmarsson et al. 2013, p. 355; Pye/Dobbins 2015, p. 12f.

[87] Thiemer/Thiemer in: Friedrichs 1997, p. 245; Donges/ Freytag 2009, p. 172.

[88] Thiemer/Thiemer in: Friedrichs 1997, p. 246.

[89] Klus 2013, p. 82; Thiemer/Thiemer in: Friedrichs 1997, p. 246; Leijten/ de Bel 2020, p. 100.

[90] Parikh et al. 2014, p. 69.

offered for sale or rent. Additionally, dwellings have relatively long production periods, meaning the housing market cannot immediately adapt to changing demand but does so with inevitable delays.[91] Once constructed, a building predetermines its emissions for many years ahead, with limited scope for significant alterations thereafter.[92]

Moreover, constructing a dwelling incurs high production costs, and the return on the initial investment is gradual due to its lengthy lifespan. This presents uncertainty for owners regarding the adequacy of their investment returns.[93] The increased energy efficiency of buildings can reduce district heating costs, potentially disincentivising investment in energy efficiency modernisation. Therefore, robust (state) policies are necessary to curtail building energy use and, consequently, greenhouse gas emissions.[94]

The dwelling can be characterised as an immobile and heterogeneous economic good. It is uniquely bound to a specific location and maintains a distinct relationship to other locations, which is a key factor influencing demanders' decisions in the housing market. This economic good also exhibits a high degree of complexity. Dwellings are not uniform; they differ based on a multitude of objective and subjective factors that affect the demander's assessment and foster competition in the housing market.[95]

From a legal perspective, the dwelling serves multiple needs. *Kreiczer-Levy* categorised these needs into four categories: "(1) the basic right to housing, (2) the home as a personal and intimate sphere that is free from the interference of others, (3) the home as the locus of familiar relations, and (4) the home as part of the neighbourhood."[96] This legal dimension adds complexity to the economic concept of the dwelling, further complicating the regulation of the (rental) housing market.

These distinctive features set the (rental) housing market apart as a unique market where the 'classical' rules of perfect competition do not fully apply. In a perfect competition scenario, demanders and suppliers mutually benefit through the free trade of goods in markets, with market prices signalling scarcity and matching supply and demand. However, due to its strong social component, the housing market is heavily regulated. This regulation includes both direct state intervention in price and quantity formation and indirect measures[97] such as housing allowances and social assistance for those in need.[98]

These unique characteristics indicate that the housing market cannot be considered an ideal-typical market. The goods traded are not entirely homogeneous, and not all market

---

[91] Klus 2013, p. 82f.

[92] Parikh et al. 2014, p. 697.

[93] Klus 2013, p. 83.

[94] Parikh et al. 2014, p. 697.

[95] Klus 2013, p. 83.

[96] Kreiczer-Levy in: Vols/Sidoli (eds.) 2017, p. 26.

[97] Thiemer/Thiemer in: Friedrichs 1997, p. 245.

[98] OECD 2020, p. 15.

Tenant T			
		Rent increase	No rent increase
	investment	I T2-2L	II T1-4L
Landlord L	non-investment	III T4-1L	IV T3-3L

Explanation of behaviour

L	1 = III	No own investment = cooperation performance of tenant in the form of higher rent
	2 = I	Cooperation = more advantageous than mutual defection
	3 = IV	Mutual defection
	4 = II	Own investment without advantages for tenant
T	1 = II	One-sided benefit from investment performance of landlord
	2 = 1	Cooperation = more advantageous than mutual defection
	3 = IV	Mutual defection
	4 = III	Own cooperation performance without benefits from the landlord's side

Fig. 3.1 Landlord-tenant-dilemma—Preferences of the actors. *Source* Own presentation based on: Rodi, p 2014. 35; Hallof 2013, p. 115f.

participants are fully informed about market activities. Furthermore, personal benefits and costs do not always align with societal costs. The main actors in the housing market are landlords and tenants. Landlords, as providers, offer usage rights to their dwellings on the housing market with the intent of making a profit. Their primary interests lie in achieving economic scarcity of rent throughout the rental period and enhancing the dwelling's quality through structural changes. Tenants, as demanders, seek the best possible dwelling at the lowest price,[99] influenced by factors such as location, shopping accessibility, public transport, and parking availability.

The state plays a crucial role in maintaining market equilibrium, defined as a situation where neither suppliers nor demanders exert undue market power to influence prices. Policies supporting energy modernisation, coupled with measures to curb price increases, aim to achieve a competitive equilibrium. In the housing market equilibrium, the plans of landlords and tenants converge at a certain rent price, where housing supply and demand coincide, clearing the market. This "Pareto-optimal" situation reflects complete competition and an efficient balance of marginal cost and benefit, where it is not possible to improve conditions for one market participant without adversely affecting another. At this equilibrium point, tenants meet their primary housing needs, though not all housing needs are satisfied, as this would only be possible at the satisfaction quantity Xs. The market equilibrium is thus established by satisfying most, but not all, housing needs (X*).[100]

In the dynamics of the housing market, each party must compensate for the rights they acquire, as well as receive remuneration for relinquishing these rights. Ideally, if rights are allocable, the actions of one party should not impact the costs or benefits of other market actors, meaning all costs and benefits of an activity are borne by its initiator. However, this is not always the case due to the presence of external effects.[101] External effects are the consequences of decisions that positively or negatively impact the satisfaction of needs of parties not involved in the decision-making or its implementation. These interests of landlords and tenants often diverge. If one party invests in energy modernisation, it inadvertently creates external effects for the other party.[102] For instance, if a landlord neglects economically viable energy improvements, they may not directly feel the repercussions, as the tenant bears the additional costs, separate from the basic rent.[103]

External effects are economically undesirable as they lead to suboptimal resource allocation—one party becomes the sole investor, while the other gains unjustified advantages. If the landlord undertakes energy modernisation without proportionately increasing the rent, the tenant benefits from reduced heating costs and enhanced comfort. Conversely, if the tenant invests in energy modernisation, the dwelling's value increases for the landlord, allowing for rent elevation without the landlord's involvement in the improvements. This situation describes the "split incentives" issue,[104] resulting in the

---

99 Ibid, p. 62f.

100 Thiemer/Thiemer in: Friedrichs 1997, p. 248.

101 Donges/Freytag 2009, p. 172.

102 Rodi 2014, p. 37; Kloepfer 2016, §5 Rn. 811.

103 Klinski 2010, p. 284.

104 Bauernhansl 2014, p. 25.; Colker in Coulson/Wang/Lipscomb 2012, p. 212; JRC 2017, p. 3;

landlord-tenant dilemma,[105] also known as the "investor-user-dilemma".[106] The landlord cannot predict the tenant's behaviour in the case of energy modernisation. He or she is at risk that the tenant terminates the rental contract, and the dwelling will be vacant. The tenant, in his/her turn, cannot predict when the landlord will increase the rent next time and whether it will still be reasonable for the tenant. Thus, strategic uncertainty makes decisions difficult for both parties, and each of them is afraid that the other party will use the advanced performance of others for its benefit and hence exploit the other party.[107]

The landlord-tenant dilemma can be formally illustrated by the classic example of the prisoner's dilemma—the game theory model (see Fig. 3.1).[108] If the tenant is uncertain about the landlord's actions, they may choose to defect to avoid the risk of rent increase. Similarly, the landlord may defect to avoid the worst-case scenario for him (II). Consequently, both parties "miss" the mutually beneficial preference (I) due to fear of scenarios II and III. Thus mutual "defecting" is rational (IV) to exclude the worst risk for each party. Neither the tenant nor the landlord succeeds in bringing about an ex-ante effective and, therefore, credible behavioural commitment.[109]

This dilemma structure indicates that the market has failed (the doctrine of market failure), suggesting the need for state intervention to internalise externalities into the market price through appropriate measures, thus reflecting the expenses of both parties.[110]

From the legal point of view, the challenges discussed are situated at the intersection of energy law, environmental law, and tenancy law. Landlords generally have two options regarding energy efficiency modernisation: maintain the status quo or invest in efficient retrofitting. Landlords are not legally obliged to modernise; they can simply pass on operating costs for heating and electricity to tenants. Consequently, higher energy consumption associated with outdated systems and insufficient retrofitting impacts the landlord only insofar as it affects the market value of their rental space.[111]

As mentioned earlier, neither landlords nor tenants have sufficient incentives to invest in thermal efficiency improvements. Tenants, not being property owners, are typically unmotivated to finance energy modernisation, lacking visibility of its benefits. Landlords, on the other hand, perceive tenants as the end consumers of energy renovation benefits and, therefore, feel justified in increasing rent post-modernisation.[112] Moreover, should the landlords decide to modernise the dwellings, they must bear the high costs

[105] Kholodilin et al.2016, p. 4.

[106] Kramer 2007, p. 283; Börstinghaus/ Krumm 2018, p. 635.

[107] More to this topic: Rodi 2014, p. 32f.

[108] Rodi 2014, p. 34.

[109] Ibid.

[110] Hallof 2013, p. 101.

[111] Kramer 2007, p. 284.

[112] Ástmarsson et al. 2013, p. 356; Karras/ Novikova 2021, p. 2.

themselves, tying up capital.[113] Both parties have different interests hiding behind the landlords and tenants as potential investors of such improvement. The landlords' rationale for investing money into such retrofitting is the potential effects energy efficiency has on the selling price of a dwelling and the generated rental income streams. However, there are also risks for the rental income or the risks of vacancy, which may be affected by the price changes. These risks are considered important determinants of investment decisions.[114] The tenants, in their turn, often face informational and financial barriers and uncertainty risks, which prevent them from investing money in such improvements.[115]

There are several indicators characterising a landlord-tenant-dilemma:

- Usually, tenants cannot evaluate the real thermal quality of a dwelling due to limited knowledge and technical understanding, as well as missing information on the efforts undertaken by the landlord to improve thermal efficiency.[116]
- The tenant's willingness to pay is influenced by the uncertainty of the length of the rental relationship. In case of an energy price increase, they often prefer to move to a more energy-efficient dwelling because of lower transaction costs.[117]
- Even if the transaction costs allowing to fully appropriate the returns of energy-saving investments to either the landlord or the tenant are incurred when concluding the rental contracts, they are claimed to be prohibitively high.[118]

Another issue boosting split incentives, particularly between landlords and primarily low-income tenants whose energy bills are publicly funded, is the "reverse" split incentive. This occurs when occupants are not responsible for their utility bills, e.g., because they are covered by the state and thereby have little or no interest in saving energy. However, such incentives may also occur under so-called "rent including heating costs", where the landlord bears all utility costs for heating, other capital and operating expenses.[119] This scheme of rent is famous for student housing in Germany. In the case of publicly covered bills, there is at least a "break-line" of reasonable costs that will be discussed in Chap. 5 in more detail. It stimulates affected tenants to meet energy-saving

[113] Kramer 2007, p. 284.

[114] Kholodilin et al. 2016, pp. 2–4; Karras/ Novikova 2021, p. 2.

[115] Kholodilin et al. 2016, p. 4.

[116] Kholodolin et al. 2016, pp. 4–5; Karras/ Novikova 2021, p. 2.

[117] Transaction costs are a) costs of undertaking a transaction, including search and information costs, bargaining costs and monitoring-enforcement costs of implementing a transaction; and b) the opportunity costs of non-fulfilment of an efficient transaction (Rao, 2003, p. xvi); Kholodolin et al. 2016, p. 4f.; Karras/ Novikova 2021, p. 2.

[118] Schleich/Gruber 2008; Kholodolin et al. 2016, pp. 4–5; Karras/ Novikova 2021, p. 2.

[119] JRC 2017, p. 3.

measures. The "rent including heating costs" does not stimulate the affected tenant to meet energy-saving measures as they do not bear energy costs.

An additional issue in the landlord-tenant dilemma, within the context of energy modernisation, concerns multi-tenant and multi-owner split incentives. This refers to situations where energy modernisation impacts buildings comprised of dwellings owned by several landlords and rented to multiple tenants, or by a single landlord and rented to several tenants.[120] This scenario, though seemingly complex, is increasingly common in practice. Particularly nowadays, a dwelling is often viewed in society as an investment and a source of additional security for future pensions. The decision-making processes in multi-owner buildings with dwellings rented to multiple tenants are even more intricate. Any energy efficiency initiatives in these buildings require the consensus of all decision-making parties. Furthermore, in multi-tenant and multi-owner buildings, the benefits and costs of energy efficiency renovations may differ from one apartment to another, creating further complications in achieving consensus.[121]

The landlord-tenant dilemma is a multifaceted phenomenon caused by numerous barriers and split incentives. Addressing these issues simultaneously is essential to overcoming them. Resolving a single barrier does not fully resolve the problem; for instance, even if landlords can credibly convey information about energy savings, this does not guarantee tenants will agree to pay rent that includes the total energy cost savings. Tenants may still opt to move to alternative residences, thereby increasing the risk of vacancies for landlords.[122] Moreover, even if tenancy law provides a high level of security, as is the case in Germany, for example, this security often benefits the tenant as a weaker party of a rental contract but not the landlord in the same volume.[123] Therefore, the landlord-tenant dilemma extends beyond just economic or legal disciplines.

Hallof emphasises that the landlord-tenant dilemma only arises when energy renovation benefits the tenant without any advantage for the landlord. If neither party benefits, for example, in cases of higher primary energy savings without cost savings for the tenant, the dilemma is non-existent.[124] However, considering the EU's climate goals, there is a risk of boosting the landlord-tenant dilemma if landlords are unilaterally obligated to bear renovation costs. Hence, policies addressing this issue must be formulated with great caution.

Legislators aiming to resolve issues concerning low-income tenants face the challenging task of balancing the interests of both tenants and landlords.[125] Landlords seek profits

---

[120] The situation when the house is owned by several landlords and is rented to one tenant is also conceivable but happens less frequently in practice.

[121] JRC 2017, p. 3f.; Karras/ Novikova 2021, p. 2.

[122] Kholodilin et al. 2016, p. 5; Karras/ Novikova 2021, p. 2.

[123] Karras/ Novikova 2021, p. 2.

[124] Hallof 2013, p. 26f.

[125] Levy 1976, p. 155f.

from properties rented to low-income households, but this can be problematic due to economic costs and legal constraints stemming from changes in the traditional relationship between these two social groups. Overburdened landlords often resort to abandoning their properties, further reducing the low-income housing stock.[126]

The main stakeholders of the landlord-tenant dilemma in the context of rental dwellings are the owners, facility managers and occupants. Nevertheless, the government is also a stakeholder, owing to its commitment to reducing greenhouse gas emissions. Moreover, energy suppliers are obliged to contribute to significant annual energy savings. However, a decrease in energy consumption implies reduced income for these companies. Despite this, it is crucial to recognise energy suppliers as stakeholders in the aforementioned dilemma and address them through relevant policies. It should not be overlooked that extensive property renovations are closely associated with the creation of new job opportunities, which can have a positive social[127] and economic impact, significantly helping the economy to recover after the Covid-19 crisis.

The literature discusses various general barriers to sustainable retrofitting and renovation. The most notable are economic and informational barriers. Economic barriers include a lack of financial incentives as life cycle perspectives. *Ástmarsson* et al. identify informational barriers as limited political awareness, a lack of overview and unified direction among main stakeholders, and insufficient knowledge of the potential for energy renovation and its priorities. Additionally, stakeholders often prefer new buildings over renovations, impacting the dearth of investments in renovation projects.[128]

From an economic point of view, there are several types of information deficiencies, broadly categorised into two groups: ignorance and uncertainty. Ignorance occurs when market participants lack sufficient information, a gap that could potentially be bridged through appropriate information gathering. On the other hand, uncertainty refers to future trends that cannot be predicted adequate certainty, even with the best possible efforts. This includes phenomena such as natural disasters and economic fluctuations. Uncertainty can also relate to the intentions of a transaction partner who might act unlawfully, even at the point of contract conclusion, a situation often referred to as "moral hazard".[129]

In the context of the landlord-tenant relationship, both landlords and tenants often lack information about existing state programmes for the energy refurbishment of dwellings in their respective countries. Tenants may not be aware of the various ways to conserve energy in their homes. Both parties face uncertainty regarding each other's intentions related to the duration of the contract. While the tenant is often legally protected against unlawful contract termination, the landlord lacks assurance that the

---

[126] Ibid, p. 156.

[127] Ástmarsson et al. 2013, pp. 356–357; Karras/ Novikova 2021, p. 2.

[128] Ástmarsson et al. 2013, p. 357.

[129] Rodi 2014, p. 45.

tenant will rent the dwelling for a long-term period, even when the contract is of indefinite duration. The specific legal and financial barriers in the respective countries will be identified and discussed in Chap. 5. The subsequent sub-chapter will focus on the significance of this topic for the countries being compared.

3.3 Relevance of Issues for Germany and Sweden

Both countries, Germany and Sweden, are highly developed welfare states,[130] yet the social phenomenon of poverty persists in these countries. Households at risk of poverty, as well as those in poverty, encounter additional challenges, such as maintaining thermal comfort in their dwellings, often due to financial constraints preventing them from affording better housing.

According to the Member State Report on Energy Poverty,[131] Germany performed better than the EU average on the population-reported indicators in 2019. A 2018 survey indicated that less than 3% of the population could not keep their home adequately warm, and 3.0% were behind on utility bills. Yet, Germany's performance in the expenditure-based indicators was poorer than the EU average, as 17.4% of households spent a high share of their income on energy expenditure.[132] At first glance, energy poverty may not seem a pressing issue in Germany.

A similar situation exists in Sweden, which also surpassed the EU average on population-based indicators in 2018: just over 2% of the Swedish population struggled to maintain adequate warmth in their homes, and a slightly higher percentage faced arrears on utility bills. Like Germany, Sweden's expenditure-based indicators were poorer than the EU average. The EPOV report suggests this may be due to energy costs not always being included in rental expenses.[133]

Nonetheless, until recently, Germany had not directly addressed energy poverty in its policies. While the social system prevents absolute poverty, it indirectly influences energy poverty issues. German social insurance benefits contribute to relatively low poverty rates but leave a significant poverty gap, as the average income of households receiving social benefits remains below the national poverty line.[134] It is expected that the share of people living at risk of poverty and social exclusion may increase after the Covid-19 crisis. With nearly half of the German population living in rented dwellings, many of which need modernisation to meet EU climate goals, addressing the landlord-tenant dilemma is more crucial than ever. An additional challenge is securing

[130] Behrendt 2002, p. 13.

[131] EPOV 2019.

[132] Ibid, p. 42.

[133] Ibid, p. 107.

[134] Behrendt 2002, p. 47.

new financial resources for the energy modernisation of rented housing for those with limited or no access to financial assets, who are particularly vulnerable. In striving to achieve climate goals, these challenges must be balanced and resolved through practical concordance.

In Sweden, there was almost no public discussion on the topic of energy poverty as the research has not explicitly focused on this topic; however, there were a number of studies that analysed closely related topics such as energy efficiency, renewable energy and energy transition. Sweden has well-developed social policies which help to keep energy poverty at low levels. There are also a few information and awareness measures addressing vulnerable households.[135] Comparing Swedish and German policies related to energy poverty and landlord-tenant-dilemma could help to design good policies for Germany that could contribute to its climate goals and obligations towards the EU.

References

'Archive:Income Poverty Statistics'. 2018. 2018. https://ec.europa.eu/eurostat/statistics-explained/index.php?title=Archive:Income_poverty_statistics.

'Armutsgefährdungsquote in Deutschland bis 2022'. n. d. Statista. Accessed 20 December 2023. https://de.statista.com/statistik/daten/studie/72188/umfrage/entwicklung-der-armutsgefaehrdungsquote-in-deutschland/.

Annadotter, Kerstin, Folke Björk et. al. 2014. Sustainable renovation strategy in the Swedish Million Homes Programme: A case study. https://www.researchgate.net/publication/283300883_Sustainable_renovation_strategy_in_the_Swedish_Million_Homes_Programme_A_case_study

Ástmarsson, Björn, Per Anker Jensen, and Esmir Maslesa. 2013. 'Sustainable Renovation of Residential Buildings and the Landlord/Tenant Dilemma'. *Energy Policy* 63 (December): 355–62. https://doi.org/10.1016/j.enpol.2013.08.046.

'Themenseite: Armut in Deutschland'. n.d. Statista. Accessed 20 December 2023. https://de.statista.com/themen/120/armut-in-deutschland/.

'Average Monthly Salary in European Union 2022 | Reinis Fischer'. n. d. Accessed 20 December 2023. https://www.reinisfischer.com/average-monthly-salary-european-union-2022.

Bartiaux, Françoise, Christophe Vandeschrick, Mithra Moezzi, and Nathalie Frogneux. 2018. 'Energy Justice, Unequal Access to Affordable Warmth, and Capability Deprivation: A Quantitative Analysis for Belgium' 225 (June).

Behrendt, Christina. 2002. *At the Margins of the Welfare State: Social Assistance and the Alleviation of Poverty in Germany, Sweden and the United Kingdom*. https://doi.org/10.4324/9781315188034.

'Belgium: Share of Tenants 2008–2017'. n. d. Statista. Accessed 20 December 2023. https://www.statista.com/statistics/544645/tenants-among-population-belgium/.

Boardman, Brenda. 1991. *Fuel Poverty: From Cold Homes to Affordable Warmth*. Belhaven Press.

Boardman, Brenda. 2010. *Fixing Fuel Poverty: Challenges and Solutions*. London. https://www.routledge.com/Fixing-Fuel-Poverty-Challenges-and-Solutions/Boardman/p/book/9781844077441.

[135] EPOV 2019, p. 109.

Börstinghaus, Ulf, and Carsten Krumm. 2018. 'Die "Herausmodernisierung" Als Neue Ordnungswidrigkeit Im Referentenentwurf Für Ein "Mietrechtsanpassungsgesetz". Eine Erste Stellungnahme Aus Amtsrichterlicher Sicht'. *Neue Zeitschrift Für Miet- Und Wohnungsrecht*, 633–40.

Bouzarovski, Stefan, and Sergio Tirado Herero. 2017. 'The Energy Divide: Integrating Energy Transitions, Regional Inequalities and Poverty Trends in the European Union'. *European Urban and Regional Studies* 24 (1): 69–86. https://doi.org/10.1177/0969776415596449.

Bouzarovski, Stefan, and Saska Petrova. 2015. 'A Global Perspective on Domestic Energy Deprivation: Overcoming the Energy Poverty–Fuel Poverty Binary'. *Energy Research & Social Science* 10 (November): 31–40. https://doi.org/10.1016/j.erss.2015.06.007.

Bundesministerium für Arbeit und Soziales. 2017. 'Life Situations in Germany. The German Federal Government's 5th Report on Poverty and Wealth'. https://www.bmas.de/SharedDocs/Downloads/EN/PDF-Publikationen/a306e-the-german-federal-governments-5th-report-on-poverty-and-wealth.pdf?__blob=publicationFile&v=1.

Bundesministerium für Wirtschaft und Energie. 2020. 'Integrierter Nationaler Energie- Und Klimaplan Gemäß Der Verordnung Des Europäischen Parlaments Und Des Rates Über Das Governance-System Für Die Energieunion Und Für Den Klimaschutz, Zur Änderung Der Richtlinie 94/22/EG, Der Richtlinie 98/70/EG, Der Richtlinie 2009/31/EG, Der Verordnung (EG) Nr. 663/2009, Der Verordnung (EG) Nr. 715/2009, Der Richtlinie 2009/73/EG, Der Richtlinie 2009/119/EG Des Rates, Der Richtlinie 2010/31/EU, Der Richtlinie 2012/27/EU, Der Richtlinie 2013/30/EU Und Der Richtlinie (EU) 2015/652 Des Rates Und Zur Aufhebung Der Verordnung (EU) Nr. 525/2013'. https://www.bmwi.de/Redaktion/DE/Downloads/I/integrierter-nationaler-energie-klimaplan.pdf?__blob=publicationFile&v=4.

'Bürgergeld ersetzt ALG II | Bundesregierung'. 2023. Die Bundesregierung informiert | Startseite. 1 January 2023. //www.bundesregierung.de/breg-de/schwerpunkte/entlastung-fuer-deutschland/buergergeld-2125010.

BVerwG, 08.12.1953 – I C 100.53. 1953. Urt. BVerwG.

Chen, Chien-fei, Xiaojing Xu, and Julia K. Day. 2017. 'Thermal Comfort or Money Saving? Exploring Intentions to Conserve Energy among Low-Income Households in the United States'. *Energy Research & Social Science* 26 (April): 61–71. https://doi.org/10.1016/j.erss.2017.01.009.

Commission Staff Working Document Country Report Sweden 2019 Including an In-Depth Review on the Prevention and Correction of Macroeconomic Imbalances Accompanying the Document Communication from the Commission to the European Parliament, the European Council, the Council, the European Central Bank and the Eurogroup 2019 European Semester: Assessment of Progress on Structural Reforms, Prevention and Correction of Macroeconomic Imbalances, and Results of in-Depth Reviews Under Regulation (Eu) No 1176/2011. 2019. https://eur-lex.europa.eu/legal-content/EN/TXT/PDF/?uri=CELEX:52019SC1026&rid=8.

Communication from the Commission to the Council, the European Parliament, the European Economic and Social Committee and the Committee of the Regions – Joint Report on Social Inclusion Summarising the Results of the Examination of the National Action Plans for Social Inclusion (2003–2005) {SEC(2003)1425}. 2003. https://eur-lex.europa.eu/legal-content/EN/TXT/?uri=celex%3A52003DC0773.

Coulson, N. Edward, Yongsheng Wang, and Clifford A. Lipscomb. 2012. *Energy Efficiency and the Future of Real Estate. Energy Efficiency and the Future of Real Estate*. https://doi.org/10.1057/978-1-137-57446-6.

Dabrowski, Martin, and Judith Wolf. 2018. *Armut und soziale Gerechtigkeit in Deutschland*. Brill | Schöningh. https://www.lehmanns.de/shop/sozialwissenschaften/43194176-9783506792440-armut-und-soziale-gerechtigkeit-in-deutschland.

Deutscher Bundestag. 2022. 'Gesetzentwurf Der Bundesregierung, Entwurf Eines Zwölften Gesetzes Zur Änderung Des Zweiten Buches Sozialgesetzbuch Und Anderer Gesetze – Einführung Eines Bürgergeldes (Bürgergeld-Gesetz)'. Drucksache 20/3873. https://dserver.bundestag.de/btd/20/038/2003873.pdf.

Directorate-General for Internal Policies. 2016. 'Energy Efficiency for Low-Income Households'. https://www.europarl.europa.eu/thinktank/en/document/IPOL_STU(2016)595339.

Donges, Jürgen B., and Andreas Freytag. 2009. *Allgemeine Wirtschaftspolitik*. 3., Überarb. u. erw. Aufl. Stuttgart: UTB, Stuttgart.

Economidou, Marina. 2014. 'Overcoming the Split Incentive Barrier in the Building Sector'. JRC Publications Repository. 17 September 2014. https://doi.org/10.2790/31513.

'Energy – European Commission'. 2023. 19 December 2023. https://energy.ec.europa.eu/index_en.

'Erhöhung Der ALG II Regelsätze Ab 2022 – Jobcenter Heidenheim'. n. d. Accessed 20 December 2023. https://www.jobcenter-heidenheim.de/aktuelles/erhohung-der-alg-ii-regelsatze-ab-2022/.

European Anti-Poverty Network. 2019. 'Poverty Report Sweden 2019'. https://www.eapn.eu/wp-content/uploads/2020/04/EAPN-PW2019-Sweden-EN-EAPN-4306.pdf.

Forsa. 2006. 'Meinungen Zum Thema Heizkosten Und Erneuerbare Energien. Telefonische Befragung von September 2006'. *Energiewirtschaftliche Tagesfragen* 08: 14–16.

Friedrichs, Jürgen, ed. 1997. *Die Städte in den 90er Jahren: demographische, ökonomische und soziale Entwicklungen*. Opladen: Westdeutscher Verl.

Grossmann, Katrin, André Schaffrin, and Christian Smigiel, eds. 2017. *Energie und soziale Ungleichheit: zur gesellschaftlichen Dimension der Energiewende in Deutschland und Europa*. Wiesbaden: Springer VS.

Hallof, Irene J. 2013. *Das Vermieter-Mieter-Dilemma bei der energetischen Gebäudesanierung: eine rechtliche und ökonomische Analyse*. Praxis und Theorie des Bau- und Immobilienrechts 20. Berlin Brüssel: Lexxion Der Juristische Verlag.

'Household Gas Prices up 17.7% in 1st Half of 2022'. n. d. Federal Statistical Office. Accessed 20 December 2023. https://www.destatis.de/EN/Press/2022/10/PE22_460_61243.html.

'Housing Costs: 10.7% of the Population Regarded as Overburdened in 2021'. n. d. Federal Statistical Office. Accessed 20 December 2023. https://www.destatis.de/EN/Press/2022/08/PE22_N054_61.html.

'Housing in Europe – House or Flat – Owning or Renting'. n. d. Housing in Europe. Accessed 20 December 2023. https://ec.europa.eu/eurostat/cache/digpub/housing/bloc-1a.html.

Hradil, Stefan. 2005. *Soziale Ungleichheit in Deutschland*. 8. Auflage. Lehrbuch. Wiesbaden: Springer Fachmedien Wiesbaden GmbH.

Karras, Jana, and Aleksandra Novikova. 2021. 'Germany. Legal Barriers to the Decarbonisation of Dwellings Occupied by Low-Income Tenants and Financial Incentives to Overcome These Barriers'. IKEM Climate Finance Week. https://www.ikem.de/wp-content/uploads/2021/05/Brief_5_Legal-barriers-to-the-decarbonisation-of-dwellings.pdf.

Kloepfer, Michael, and Rico David Neugärtner. 2016. *Umweltrecht*. 4. Auflage. München: C.H. Beck.

Klus, Sebastian. 2013. *Die europäische Stadt unter Privatisierungsdruck: Konflikte um den Verkauf kommunaler Wohnungsbestände in Freiburg*. Stadt, Raum und Gesellschaft. Wiesbaden: Springer VS.

Koch, Hans-Joachim. 2011. 'Koch: Klimaschutzrecht'. *Neue Zeitschrift Für Verwaltungsrecht*, no. 11: 641–54.

Kolokotsa, Denia, and Mat Santamouris. 2015. 'Review of the Indoor Environmental Quality and Energy Consumption Studies for Low Income Households in Europe'. *Science of The Total Environment* 536 (December): 316–30. https://doi.org/10.1016/j.scitotenv.2015.07.073.

Kholodilin, Konstantin, Andreas Mense, and Claus Michelsen. 2016. 'The Market Value of Energy Efficiency in Buildings and the Mode of Tenure'. Urban Studies 54 (September). https://doi.org/10.1177/0042098016669464.

Kramer, Dennis R. 2007. 'Kramer: Energieeinsparung Im Mietwohnsektor Durch Wärme-Contracting'. *ZUR*, no. 6: 283–89.

Leal-Arcas, Rafael, and Jan Wouters, eds. 2017. *Research Handbook on EU Energy Law and Policy*. Research Handbooks in European Law. Cheltenham, UK: Edward Elgar Publishing.

Leijten, Ingrid, and Kaisa Bel. 2020. 'Facing Financialization in the Housing Sector: A Human Right to Adequate Housing for All'. *Netherlands Quarterly of Human Rights* 38 (June): 94–114. https://doi.org/10.1177/0924051920923855.

Levy, Mont S. 1976. 'Adjusting the Economic Relationship of Landlord and Tenant—Rent Alteration Remedies'. *Urban Law Annual* 11 (1). https://journals.library.wustl.edu/urbanlaw/article/id/8295/.

Mudambi, Ram. 2004. 'The Economics of Transaction Costs, Edited by Rao, P.K., Houndmills, Basingstoke, Hampshire, UK and New York: Palgrave Macmillan, 2003, Xvi+197 Pp., GBP 50.00(Cloth)'. *Managerial and Decision Economics* 25 (January): 56–57. https://doi.org/10.1002/mde.1140.

Odekon, Mehmet A., ed. 2015. *The SAGE Encyclopedia of World Poverty*. Second Edition. Thousand Oaks, California: SAGE Publications, Inc.

OECD. 2020. 'Housing Policies for Sustainable and Inclusive Cities: How National Governments Can Deliver Affordable Housing and Compact Urban Development'. Regional Development Working Papers 2002/03. https://www.oecd-ilibrary.org/docserver/d63e9434-en.pdf?expires=1637766762&id=id&accname=guest&checksum=8B05FE6EE297597E7D39F2B41B2ECE21.

Parikh, Jyoti et al. '"Buildings". In: Climate Change 2014: Mitigation of Climate Change. Contribution of Working Group III to the Fifth Assessment Report of the Intergovernmental Panel on Climate Change'. In . https://www.ipcc.ch/site/assets/uploads/2018/02/ipcc_wg3_ar5_chapter9.pdf.

People and Buildings: Comparative Housing Law. 2017. Studies in Housing Law. The Hague: Eleven International Publishing.

Podgornik, Ales, Boris Sucic, and Bostjan Blazic. 2016. 'Effects of Customized Consumption Feedback on Energy Efficient Behaviour in Low-Income Households'. *Journal of Cleaner Production* 130 (February). https://doi.org/10.1016/j.jclepro.2016.02.009.

Pye, Steve, Audrey Dobbins, Claire Baffert, Jurica Brajkovic, Rocco Miglio, and Paul Deane. 2015. *Energy Poverty and Vulnerable Consumers in the Energy Sector across the EU: Analysis of Policies and Measures*.

'Regelsätze steigen zum 1. January 2021'. 2021. Die Bundesregierung informiert | Startseite. 1 January 2021. //www.bundesregierung.de/breg-de/service/archiv/regelsaetze-steigen-1775798.

REPORT FROM THE COMMISSION TO THE EUROPEAN PARLIAMENT, THE COUNCIL, THE EUROPEAN ECONOMIC AND SOCIAL COMMITTEE AND THE COMMITTEE OF THE REGIONS 2020 Report on the State of the Energy Union Pursuant to Regulation (EU) 2018/1999 on Governance of the Energy Union and Climate Action. 2020. https://eur-lex.europa.eu/legal-content/EN/TXT/?uri=COM%3A2020%3A950%3AFIN.

Rodi, Michael. 2014. Ökonomische Analyse des Öffentlichen Rechts. Lehrbuch. Berlin Heidelberg: Springer Gabler.

Schleich, Joachim, and Edelgard Gruber. 2008. 'Beyond Case Studies: Barriers to Energy Efficiency in Commerce and the Services Sector'. *Energy Economics* 30 (March): 449–64. https://doi.org/10.1016/j.eneco.2006.08.004.

Schreiner, Nadine. 2015. 'Auf Der Suche Nach Energiearmut: Eine Potentialanalyse Des Low-Income-High-Costs Indikators Für Deutschland (Searching for Fuel Poverty – A Potential Analysis of the Low-Income-High-Costs Indicator for Germany)'. SSRN Scholarly Paper. Rochester, NY. https://doi.org/10.2139/ssrn.2710713.

Statbel. n. d. 'More than 2 Million Belgians at Risk of Poverty or Social Exclusion'. P – Armoederisico's 2019. Accessed 20 December 2023. https://statbel.fgov.be/en/themes/households/poverty-and-living-conditions/risk-poverty-or-social-exclusion.

Thema, Johannes, and Florin Vondung. 2020. 'EPOV Indicator Dashboard. Methodology Guidebook'. Wuppertal Institut für Klima, Umwelt, Energie GmbH. https://energy-poverty.ec.europa.eu/system/files/2021-09/epov_methodology_guidebook_1.pdf.

Verbraucherzentrale Nordrhein-Westfalen. 2015. 'Gemeinsame Wege Aus Der Energiearmut'. Düsseldorf. https://www.verbraucherzentrale.nrw/sites/default/files/migration_files/media237456A.pdf.

Bauernhansl, Thomas, ed. 2014. Energieeffizienz in Deutschland - Eine Metastudie. Berlin/Heidelberg.

Commission Recommendation (EU) 2020/1563 of 14 October 2020 on energy poverty. (2020). In OJ L (Vol. 357). http://data.europa.eu/eli/reco/2020/1563/oj/eng

Commission Recommendation (EU) 2020/1563 of 14 October 2020 on energy poverty. (2020). In OJ L (Vol. 357). http://data.europa.eu/eli/reco/2020/1563/oj/eng

Communication from the Commission to the European Parliament, the Council, the European Economic and Social Committee and the Committee of the Regions 'Fit for 55': Delivering the Eu's 2030 Climate Target on the Way to Climate Neutrality (2021). https://eur-lex.europa.eu/legal-content/EN/ALL/?uri=COM%3A2021%3A550%3AFIN

Income and Wealth Poverty in Germany SSRN Electronic Journal https://doi.org/10.2139/ssrn.2833244

Joint Research Centre of European Commission (JRC). 2017. Overcoming the split incentive barrier in the building sector. Unlocking the energy efficiency potential in the rental & multifamily sectors. http://publications.jrc.ec.europa.eu/repository/bitstream/JRC101251/ldna28058enn.pdf

Klinski, Stefan. 2010. Energetische Gebäudesanierung und Mietrecht – Hemmnisse und Reformüberlegungen. ZUR 2010, 283–290.

Report from the Commission to the European Parliament, the Council, the European Economic and Social Committee and the Committee of the Regions 2020 Report on the State of the Energy Union Pursuant to Regulation (Eu) 2018/1999 on Governance of the Energy Union and Climate Action (2020). https://eur-lex.europa.eu/legal-content/EN/TXT/?uri=COM%3A2020%3A950%3AFIN

Richtlinie 2010/31/EU des Europäischen Parlaments und des Rates vom 19. Mai 2010 über die Gesamtenergieeffizienz von Gebäuden (Neufassung), EP, CONSIL, 153 OJ L (2010). https://data.europa.eu/eli/dir/2010/31/oj/deu

EU Regulative Framework

4

4.1 Legal Policy Background

Before delving into the analysis of the European policy framework, it is important to emphasise that the legal aspects of decarbonising dwellings occupied by low-income households represent a multifaceted phenomenon encompassing numerous branches of law, including climate law, energy and energy efficiency law, social law, rental law, etc. European law often adopts a one-size-fits-all approach, contrasting with the diverse and significantly different legal orders of member states in addressing similar phenomena. This contradiction inhibits the effectiveness of multipolar steering instruments,[1] while simultaneously establishing a framework for legal orders by setting the intended direction. Considering the competencies of the EU, this alignment cannot always occur imperatively but often demands substantial political will from the member states themselves. Understanding the European regulatory framework's role in the landlord-tenant dilemma aids in recognising the existing boundaries and potential flexibilities within national orders for addressing this issue. This chapter will analyse the principal regulatory aspects related to the topic of this research.

Despite the importance of European energy efficiency policy in setting the framework for the issue at hand, the average practising lawyer usually engages more with national rather than European laws and policies. This situation arises for two primary reasons. Firstly, it reflects the national sovereignty of member states, where the national legislator, directly legitimised by the citizens, confronts them with legal norms. Secondly, for

[1] Hartmann 2011, p. 636.

J. Karras, *Legal Barriers to the Energy Modernisation of Dwellings Occupied by Low-income Tenants and Opportunities to Overcome these Barriers*, Schriftenreihe des Instituts für Klimaschutz, Energie und Mobilität,
https://doi.org/10.1007/978-3-658-44193-7_4

most cases, European primary law stipulates a two-stage legislative process where the European “framework law” precedes the national implementing law. Concurrently, in the realm of energy efficiency law, directly applicable regulations that do not necessitate further implementation are scarce.[2] This makes the analysing of the European regulative framework essential as it has a significant impact on the national laws. A decade ago, approximately 80% of national environmental laws were influenced, to some extent, by EU law.[3] Presently, it is reasonable to assert that this figure is close to 100%.

The decarbonisation of the residential stock has a direct impact on CO_2 reduction and thus, on climate mitigation in general. Consequently, climate policy forms a comprehensive framework for the research topic, establishing fundamental principles and an overarching direction for actions in this domain.[4] Within the European2015 context of climate policy implementation, the national and sub-national levels play a pivotal role. This is because the EU itself contributes minimally to greenhouse gas emissions, thus possessing a limited carbon footprint.[5]

In recent years, awareness of climate mitigation has evolved at the supranational level. It is now clear that the atmosphere is warming, affecting the entire planet Earth and its inhabitants. According to Special Eurobarometer 490, 95% of EU citizens perceive climate change as a serious issue, with a substantial majority advocating for heightened responses to its challenges.[6] The Paris Agreement of December 2015 marked a new chapter in international climate mitigation law. The Intergovernmental Panel on Climate Change (IPCC) Special Report on the impacts of global warming of 1.5°C above pre-industrial levels along with related global greenhouse gas emission pathways, confirms that the effects of climate change escalate swiftly with the increase in global mean temperature. These effects become irreversible and dramatic at a 2°C rise.[7] This understanding has necessitated a re-evaluation of the importance of decarbonising the housing sector, particularly concerning its proportion of CO_2 emissions. Consequently, this has led to the development of a regulatory framework at all levels.

In the light of the rising awareness of climate mitigation and the threats posed by climate change, particularly global warming, a political reorientation acknowledging ecological limits has become imperative. The German government has identified planetary boundaries as definitive constraints for political decision-making.[8] As a part of the international community, the EU reacted to numerous climate change challenges with several

---

[2] Meßerschmidt 2011, §2 Rn. 353f.

[3] Ibid, §2 Rn. 356.

[4] Topp in Theobald/Kühling 2021, Energierecht, Gebäuderichtlinie, Rn. 78.

[5] Woerdman/ Roggenkamp/ Holwerda , p. 274.

[6] European Commission. 2019.

[7] IPCC Special report 2019; COM (2020a) 80 final; Calliess/Dross 2020, p. 457.

[8] Bundesregierung 2016, S. 12; Calliess/ Dross 2020, p. 456.

measures. The European Parliament passed on 14 March 2019 a resolution on climate change.[9] Several months later, in another resolution, the Parliament emphasised that the Union, alongside other major global economies, should commit to achieving net-zero greenhouse gas emissions as soon as feasible, but no later than 2050.[10] The European Parliament has also implored the Commission to assess the climate and environmental impact of all legislative or budgetary proposals, ensuring their compatibility with the goal of limiting global warming to below 1.5°C, and to avoid contributing to biodiversity loss.[11] Furthermore, the European Parliament called for the necessary transition to a climate-neutral society by 2050 at the latest.[12] These initiatives signify that both European and national regulatory frameworks pertaining to the decarbonisation of economies are continuously being reviewed and developed.

The landlord-tenant dilemma is tightly related to the general decarbonisation of the residential stock. The regulative framework governing this includes the European Green Deal,[13] the Renovation Wave,[14] the European Treaties, and specific directives and regulations.[15] Their relevance for the researched topic will be elaborated upon in subsequent sections. Other elements of this framework, such as the Smart Readiness Indicator,[16] the 2030 Climate target plan,[17] EU Energy Strategy,[18] Annual Sustainable Growth Strategy,[19] Circular Economy Action Plan,[20] Heating and Cooling Strategy,[21] Regulation establishing the Recovery and Resilience facility,[22] will not be analysed in depth. Although these are integral to the Renovation Wave initiative of the European Green Deal, they hold limited direct relevance to this research.

[9] European Parliament resolution 2019

[10] European Parliament resolution 2019.

[11] COM (2020a, b) 80 final.

[12] European Parliament resolution 2020.

[13] COM (2019) 640 final.

[14] COM (2020) 662 final.

[15] Hermelink/Bettgenhäuser 2021, p. 9f.

[16] 'Smart Readiness Indicator - European Commission (n.d).

[17] COM (2020) 562 final.

[18] 'Energy Strategy' (n.d).

[19] COM (2020) 575 final.

[20] COM (2020) 98 final.

[21] COM (2016) 51 final.

[22] Regulation (EU) 2021/241.

The decarbonisation of dwellings is one of the instruments of climate mitigation,[23] given that the building sector accounts for 40% of the EU's energy consumption and 36% of its greenhouse gas emissions.[24] This sector is, however, also the key to climate change mitigation policies.[25] Around 75% of the European building stock is energy inefficient.[26] The residential sector has been identified as one of the "key sectors", necessitating a multitude of actions to enhance energy efficiency at all levels—European, national, regional, and local.[27]

When evaluating the EU energy policy, it is crucial to recognise that it is founded on the principle of solidarity. This principle, while fostering unity, can complicate the establishment of a uniform and effective policy at the EU level. The EU, as a supranational organisation,[28] encompasses 27 member states with varied economic statuses, sizes, and climates. Consequently, it is unreasonable to expect uniformity in energy efficiency policies and goals across all member states. Neither is it feasible to anticipate that some members set highly ambitious objectives while others contribute minimally towards achieving net-zero emissions. This principle is further reinforced by the concept of regional differentiation, as enshrined in Article 191(2) of the Treaty on the Functioning of the European Union (TFEU). This acknowledges the vast disparities among member states and their regions in terms of economic development, geographical location, population density, ecological awareness, and environmental conditions.[29] However, the diversity of member states often precludes the adoption of a one-size-fits-all approach to solutions.[30] One of the challenges is to find a balance between the given framework and the leeway of the member states when designing solutions for the researched problem. While the member states must implement the given frameworks, they still have enough leeway in designing but also in implementing tougher measures than those proposed at the EU level for all members.

Moreover, the role of EU law in comparative analysis cannot be overstated. Examining the EU legal framework provides deeper insights into the interplay between supranational and national levels.[31] For legal researchers exploring foreign legal systems,

---

[23] Topp in Theobald/Kühling 2021, Energierecht, Gebäuderichtlinie, Rn. 78.

[24] 'Energy Efficiency in Buildings' (n. d).

[25] Parikh et al. (2014) p. 677.

[26] 'Energy Efficiency in Buildings' (n. d).

[27] European Commission (2021) 69 final. Koch 2011, p. 643.

[28] Oppermann/Classen/Nettelsheim 2016, §4 Rn. 8.

[29] Calliess in: Calliess/Ruffert 2016, Art. 191 AEUV Rn. 22.

[30] Ibid.

[31] Engelbrekt 2015, p. 88.

this serves as a vital tool for understanding another country's legal order by comparing it with their own and identifying potential solutions. This approach also furnishes researchers with the necessary legal context for subsequent comparative analyses.

In summary, comprehending the EU's framework for climate mitigation and energy efficiency is imperative for addressing the landlord-tenant dilemma. It highlights the parameters within which solutions must be sought while also delineating the scope for member states to independently implement (and potentially exceed) EU-level solutions.

4.1.1 European Policies as Guidelines for Future Legislation

4.1.1.1 European Green Deal as a Milestone in EU Climate Change Mitigation Policy

The European Commission uses the Communications instrument to provide guidelines for the European Union and its member states on various issues. A significant development in EU climate change mitigation and energy efficiency policymaking is the European Green Deal (EGD), presented as a communication. The EGD is intended as a roadmap[32] to steer the EU's economy towards greater sustainability, with the

[32] The Commission's communications are typically categorised as 'soft law' within the EU, indicating that they do not have binding force on individuals, but may still produce specific legal effects. However, the binding nature of such communications, particularly the EGD, is a subject of debate among scholars. For instance, Sikora argues that the binding force of the EGD is derived from Articles 192(3) and 191(2) of TFEU. He suggests that environmental action programmes, as outlined in these articles, possess legal status within the EU's environmental law framework. Under the Art. 192 (3), general action programmes setting out priority objectives to be attained are adopted following the ordinary legislative procedure. Thus, such programmes are legally binding for the EU institutions and are not soft law but a "normal", hard law. Moreover, Sikora posits that the principles underpinning the EGD are those environmental principles specified in Article 191(2) TFEU, which mandates their observance by both EU institutions and member states in implementing EU law and developing policies. Nonetheless, member states retain the right to introduce more stringent protective measures in accordance with the Treaties, as per Article 193 TFEU. See Sikora 2021, p. 688f.

Conversely, arguments against the direct binding force of the EGD highlight Article 288 TFEU, which does not recognise communications as a legally binding source of law. Asserting such binding force is deemed problematic as it could infringe upon the distribution of competencies within the EU. The Union's competencies are limited to those conferred by member states through the founding treaties (Article 4(1) and Article 5(2) s. 2 TEU). The creation of new legal sources with direct binding force without treaty consent from member states is considered an overreach of the Union's conferred competencies. See: Herdegen 2020, §8 Rn. 74 f; Hobe/Fremuth 2020, §7 Rn. 30f.; Horspool/ Humphreys/ Wells-Greco 2018, p. 130.

For the legally binding force of the EGD to be established, it is not sufficient to aim merely at legal effects. The EGD must aim at new or independent legal effects not already contained in the provisions of TFEU. See: Senden, p. 259. It also does not fall under the "Implied Powers" doctrine, according to which the Union can, within narrow limits. This is because the TFEU has given intermediate goal of making Europe the first climate-neutral continent.[33] Hence, the EGD is a broad-based environmental protection programme, primarily rooted in the cross-cutting clause of Article 11 TFEU.[34] It proposes multiple measures and instruments to reduce greenhouse gas emissions, preserve and restore ecosystems and biodiversity, and foster a toxic-free environment, aiming to mainstream sustainability across all EU policies.[35]

This research focuses on the EGD due to its attention to energy poverty. It underscores the need to support households unable to afford essential energy services, thereby highlighting the issue of energy poverty. The recognition of regulatory barriers for investment mobilisation for energy modernisation of rented and multi-ownership buildings is at the same time acknowledgement of the split incentives in this sphere. It is, however, not the first time when the split incentives and the landlord-tenant dilemma have been addressed at the EU level.[36] The urgency of the problem is accentuated now more than ever with the EU's enhanced decarbonisation targets, elevating the importance of this research topic. According to the EGD, around 50 million consumers struggle to adequately heat their homes. The EGD also stresses the need for socially equitable transition policies at both EU and national levels, with affordable solutions for those impacted by carbon pricing policies.[37]

The previous EU-wide greenhouse gas reduction target has to be risen from 40% to 55% compared to 1990 by 2030to achieve this new ambitious goal,[38]as the previous goal is insufficient for achieving a gradual transition to a climate-neutral economy by 2050.[39] Considering the CO_2 emission amount being produced every year by the residential

---

precise contours to the Union's competencies related to law-making. These exercise competencies are not expressly assigned to it if their use follows a certain internal logic from expressly conferred powers. See: Herdegen 2020, §8, Rn. 75f.; Hobe/Fremuth 2020, §7 Rn. 59f..; Horspool/ Humphreys/ Wells-Greco 2018, p. 132. Furthermore, this does not fall under the "general power" to act where none yet exists of Art. 352 TFEU (so-called "*flexibility clause*") because it is not evident why such a recognition of the Commission's communication as one of the legal sources of "hard" law is necessary to realise one of the Treaty objectives. See: Herdegen 2020, §8 Rn. 78; Hobe/Fremut 2020, §7 Rn. 52f.

[33] COM (2019) 640 final; Sikora 2021, p. 687.

[34] Germelmann 2020, p. 377; Calliess/ Dross 2020, p. 458.

[35] COM (2019) 640 final; Sikora 2021, p. 687; Calliess/Dross 2020, p. 457.

[36] The landlord-tenant dilemma has been described and addressed in the numerous EU acts, e. g. Art. 4 (1) lit. a) of the Energy Efficiency Directive 2012/27/EU.

[37] COM (2019) 640 final.

[38] As set by the European Council, "Framework for Climate and Energy Policies to 2030" of 23/24.10.2014, EUCO 169/14; Schlacke/Köster/Thierjung 2021a, p. 621; Schlacke/Köster/ Wentzien/Thierjung 2021b, p. 7.

[39] SWD (2020) 176 final (Part 1 + 2).

building sector, it becomes evident that urgent energy efficiency measures have to be implemented in this sector. Otherwise, it will be impossible to achieve this ambitious greenhouse reduction goal. This direction is further solidified by the EU Climate Law, initiated by the EGD.[40]

Moreover, the EGD marks a milestone in promoting a socially just transition. The increased EU-wide greenhouse gas reduction target directly influences policies concerning building sector decarbonisation, a central theme of the EGD. This is not only due to its alignment with the zero-net emission goal but also due to its connection with energy poverty.[41]

By June 2021, the Commission was expected to review and propose revisions to all relevant climate-related policy instruments, including potentially extending European emission trading to new sectors, such as the building sector. Additionally, the Commission was mandated to adopt a more ambitious EU strategy on adapting to climate change as per the EGD.[42]

Financing is a key aspect of achieving these ambitious targets, and it is central to this research. The EGD recognises the need for diverse financing mechanisms, adjustments to the regulatory framework, and additional measures for the energy transition.[43] The EGD's financial instrument estimates a requirement of at least one trillion euros over ten years for the EU's climate protection transformation. According to current estimates, tightening climate protection measures by 2040 will require even more investment. Considerable sums of money will be required in the nearest future not only to implement climate protection measures but also to compensate for damage and finance adaptation measures. Thus, it is consistent to refer to alternative resources of financing,[44] as the burdens related to all these measures are too high for the public funds.

This research will focus on mobilising private investment, estimated at €260 billion in additional annual investment. The role of existing financial funds of the EU shall be reviewed.[45] The EGD explicitly mentions the private investors as the public financial sources will not be sufficient for the energy transition.[46] The EGD's 'Investment Plan for the European Green Deal' reiterates the importance of private investment.[47] Therefore, solutions developed in this research will primarily aim to overcome regulatory barriers in mobilising private investment in the decarbonisation of the building sector.

---

[40] Schlacke/Köster/Wentzien/Thierjung 2021b, p. 7.

[41] COM (2019) 640 final.

[42] Ibid.

[43] Germelmann 2020, p. 376; Calliess/Dross 2020, p. 458.

[44] Germelmann 2020, p. 376; Moczko 2020, p. 196.

[45] COM (2019) 640 final.

[46] Germelmann 2020, p. 378.

[47] COM (2019) 640 final.; Germelmann 2020, p. 378.

The EGD plays a significant role in the energy-efficient renovation of the existing building stock, particularly of the low-income households, as the EU strives for a sustainable transformation of its economy and society. The EU's coordinated international efforts towards a coherent financial system supporting sustainable solutions also include promoting sustainable and inclusive growth,[48] which is impossible without improving the thermal living conditions of low-income tenants.

The subsequent "Fit for 55" legislative package bolsters the EGD.[49] This package, alongside the European Green Deal Communication,[50] launches a new growth strategy for the EU and its members, aiming for a fair, prosperous society with a modern, resource-efficient, competitive economy, and zero greenhouse gas emissions by 2050. This strategy seeks to decouple economic growth from resource use. The elevates the Commission's goal of making Europe the first climate-neutral continent by 2050[51] and introduces new challenges for the energy modernisation of the residential building stock, particularly in terms of financing. The legal standing of the EGD (and other Communications) will be explored in greater detail in Sub-chapter 4.4.

4.1.1.2 Renovation Wave

The EGD has catalysed the Renovation Wave (RWI), [52] aimed at enhancing building renovations for both climate neutrality and recovery, while also facilitating faster and deeper renovations for improved buildings.[53] A critical aspect of the RWI is to increase the current annual renovation rate of the building stock, which varies between 0.4% and 1.2% across member states. To meet the EU's energy efficiency and climate objectives, this rate must be at least doubled.[54] The RWI aims to promote significant energy and resource efficiency improvements in both private and public buildings.[55]

This initiative serves as an additional impetus for member states to devise socially acceptable solutions to the landlord-tenant dilemma. It is envisaged that legislation on the energy performance of buildings will be consistently enforced, starting with an assessment of Member States' long-term national renovation strategies in 2020.[56] All in all, renovations must become more attractive for both public and private sectors through stricter regulations, standards and information related to the energy performance of buildings. Member states are encouraged to provide accessible and targeted financing,

---

[48] COM (2019) 640 final; Calliess/Dross 2020, p. 458.

[49] COM (2021) 550 final; Schlacke/Köster/Thierjung 2021a, p. 621.

[50] COM (2019) 640 final.

[51] COM (2020a, b) 80 final.

[52] COM (2020) 662 final.

[53] COM (2020) 662 final; Calliess/Dross 2020, p. 459.

[54] COM (2019) 640 final; Falke 2021, p. 59.

[55] Falke 2021, p. 59; Karras/ Novikova 2021, p. 1.

[56] Calliess/ Dross 2020, p. 459.

as well as capacity building for the preparation and implementation of renovation projects. Additionally, there is a focus on developing district-based approaches that integrate renewable energy and digital solutions, aiming to create districts with a balanced energy footprint.[57]

The RWI addresses the financial barriers to renovation, highlighting a need for simple, attractive, and easily accessible public incentives, alongside mainstream financing products. Emphasising the better use of EU and national public funds and the mobilisation of private investment, the RWI considers green loan and mortgage financing as promising financial schemes. Moreover, the RWI acknowledges the link between energy poverty and inefficient buildings,[58] thereby setting a framework for addressing the landlord-tenant dilemma within the scope of this research.

4.1.1.3 "Fit for 55" Legislative Package

The European Commission's "Fit for 55% legislative package" represents a crucial milestone in EU policy, particularly concerning energy poverty. This package comprises a set of interconnected proposals designed to sharpen existing legislation, facilitate a fair, competitive, and green transition by 2030, and support the EU's recovery from the Covid-19 pandemic.[59] The economic shutdown had an enormous influence on energy poverty risk. Many people have been put on short-time work, many companies have gone bankrupt, and people have lost their jobs and thus their income and have been threatened by subsistence fear. The post-pandemic recovery is highly correlated with energy poverty as low-income households have suffered from the economic crisis and the rising energy prices,[60] more than households with a middle or high income. At the same time, it was expected that the share of energy poor people becomes higher due to the economic losses.

The "Fit for 55" package aims to support a socially equitable transition, mitigating hardships for lower-income groups and ensuring transparent communication of distribution effects. It is a step towards enhanced cooperation in climate protection[61] and addresses energy poverty more vigorously. The European Commission, in its communication related to the "Fit for 55" package, emphasises that combating pre-existing inequalities is not only a matter of fairness but a broader societal necessity.[62] This is a clear signal for the member states to take an action and ensure the just transition. However, there are disparities in national approaches. For example, Germany's Integrated National

---

[57] Falke 2021, p. 59.

[58] COM (2020b) 662 final; Hermelink/ Bettgenhäuser 2021, p. 12f; Falke 2021, p. 59; Karras/ Novikova 2021, p. 1.

[59] COM (2021) 550 final.

[60] COM (2021) 660 final.

[61] Löschel et al. 2021; Hummel/ Kolb, 2021.

[62] COM (2021a, b) 550 final.

Energy and Climate Plan (NECP) suggests that its national energy and social laws provide a comprehensive framework to cover minimum subsistence levels, including energy costs,[63] implying a non-existence of energy poverty. This stance overlooks individuals ineligible for social assistance, leaving a significant portion of the population unaddressed in national energy poverty policies.

The "Fit for 55" legislative package proposes the establishment of a new Social Climate Fund to support a fair transition and alleviate social hardships.[64] It suggests to provide financing to the member states for investment in energy efficiency, new heating and cooling systems, and clearer mobility.[65] A quarter of the revenues from the second emissions trading system shall be transferred to this fund,[66] supplemented by contributions from the member states at least 50%.[67]

All the above-mentioned stresses the importance of the researched topic. On the one hand, the EU Commission has signalised to recognise the importance of the problem and creates additional supportive mechanisms to ensure the just transition. On the other hand, the EU Commission is ambivalent strengthening in the EGD that public funds cannot be the only one source financing the just transition (including the energy modernisation of the residential building stock occupied by low-income tenants). The mobilisation of private investment shall be one of the central tasks at all levels.

4.1.1.4 REPowerEU Plan

The REPowerEU Plan stands as a pivotal recent policy document in the context of this research.[68] Russia's military aggression against Ukraine that impacted the whole world's energy system has recently challenged the European Union and interrogated its climate goals. A quick and efficient action was needed to reduce the EU dependency on Russian gas, oil and coal imports.[69] The energy crisis provoked by Russian aggression and the following political actions particularly concerns the vulnerable social groups for whom the increasing energy prices might become an unacceptable burden. At the same time,

[63] NECP Germany 2022, p. 58.

[64] COM (2021) 550 final; Löschel et al. 2021.

[65] EU Economy and Society to Meet Climate Ambitions n. d; Löschel et al. 2021.

[66] The existing emission trading system, which is regulated by the Directive 2003/87/EC of the European Parliament and of the Council of 13 October 2003, established a system for greenhouse gas emission allowance trading within the Union and amended Council Directive 96/61/EC, OJ L 275 25.10.2003, p. 32f.. (with amendments), focuses on GHG emissions for power generation and energy-intensive industry. A new emission trading scheme will be established for the transport and building sectors, Löschel et al. 2021.

[67] COM (2021) 550 final.

[68] COM (2022) 230 final.

[69] COM (2022) 230 final.

the EU should not leave the sustainable path of energy transition while overcoming new challenges The REPowerEU Plan is a strategic response, aiming to rapidly diminish the EU's dependence on Russian fossil fuels by accelerating the clean energy transition and bolstering cooperation for a more resilient energy system.[70] This plan also serves as a directive to member states, urging them to continue with just transition pathways and to keep residential building stock modernisation programmes on course. Given the new geopolitical context, the development of new financial instruments to support energy modernisation, thereby facilitating energy savings, is vital for all EU members. This is particularly crucial for countries still heavily reliant on Russian fossil fuels, including Germany.[71]

Energy savings are one of the pillars of the REPowerEU Plan. By reducing energy consumption, the EU can decrease its dependence on Russian fossil imports. Simultaneously, enhancing energy efficiency is a key aspect of these savings, contributing significantly to the clean energy transition and strengthening the resilience of the EU's economy. The new plan foresees the update of gas consumption goals set by the "Fit for 55" package and energy reduction consumption planned by the Energy Efficiency Directive proposal.[72] The REPowerEU Plan encourages the Parliament and Council to enable additional savings and energy efficiency in buildings through the EPBD.[73]. The relevance and urgency of the researched topic in this evolving geopolitical and energy context are thus clearly underscored.

4.2 Treaty on European Union and Treaty on Functioning of European Union as EU Primary Law

An in-depth examination of the EU legal framework for energy efficiency is crucial to understanding the Union's objectives and the boundaries of its competencies in this area. This will later help to decide on which level the amendments have to be taken to solve the landlord-tenant dilemma. With this respect, the following sub-chapter will analyse the legal framework of EU primary law considering the researched subject. A key functional question for the comparative analysis in this research is whether the legal systems under investigation ensure a certain standard of living conditions for their citizens, including thermal energy efficiency. This aspect is particularly relevant in the context of ensuring adequate heating and energy-efficient homes. The exploration of this question will commence with establishing the legislative framework of the researched topic at the level of EU primary law. Following this, the analysis will seek to ascertain whether such

[70] Ibid.

[71] Bingener 2022.

[72] COM (2021) 558 final.

[73] COM (2022) 230 final.

a guarantee of living conditions, with a specific focus on thermal energy efficiency, is guaranteed at this fundamental legal level.

4.2.1 Objectives

The policies are the guidelines for the further legislative procedure; they set up the trajectory for the legislator and help form expectations for the community. The primary sources of law set the limits for the EU organs to take action within this trajectory. This is the reason for briefly analysing these limits within the primary sources of EU law, which is a current constitutional framework for all EU policies,[74] including decarbonisation of dwellings and energy poverty.

At this level, three documents shall be mentioned– the Treaty on European Union (TEU), the Treaty on the Functioning of the European Union (TFEU), and the Charter of Fundamental Rights (CFR). These legal sources set the limits and anchor the main principles and the general trajectory for further legislative development. These limits shall be considered later when designing policy proposals for tackling the landlord-tenant dilemma.

In forming the concrete policy and decarbonisation stage, the principle of solidarity is one of the guiding ones. It is anchored in Art. 3 (3) TEU[75] and is a fundamental principle for energy and climate policy as well.[76] In this context, solidarity should be understood as an indicator of these interdependencies and interconnections between previously separate national markets. One country's actions and policies will affect the other EU countries. Moreover, solidarity reference attempts to create a corrective mechanism when the number of member states increases and becomes more differentiated.[77] The legal character of the principles is beyond the scope of this dissertation because the prevailing opinion, which is also represented in this thesis, affirms their directive power to steer the policy and, at the same time, sufficient flexibility to keep the decision-making space open.[78]

Numerous EU aims and objectives anchored in primary law are pertinent to the decarbonisation of the residential sector and the fight against energy poverty. They have been further concretised in secondary law sources. The definition of objectives in Art. 3 TEU

[74] Woerdman/ Roggenkamp/ Holwerda 2015, p. 279.

[75] Oppermann/Classen/Nettelsheim 2016, §4 Rn.23.

[76] Art. 2 (2), art. 6 (3) of the ECL;

[77] Leal-Arcas/ Wouters in: Leal-Arcas/ Wouters 2017, p. 18f; Woerdmann/ Roggenkamp/ Holwerda 2015, p. 277; Dagilyte in: Biondi/ Dagilyte/ Kücük 2018, p. 71f.

[78] Winter in Jans 2003, p. 10; Meßerschmidt 2011, §3 Rn. 70f.; Becker in: Becker et al. 2019, Art. 3 EUV Rn. 5; Terhechte in: Grabitz/Hilf/Nettesheim 2021, Art. 3 EUV Rn. 13.

has fundamental importance for the EU as all actions must ultimately be based thereon.[79] They leave, however, considerable space for flexibility of actions. The setting and amendment of the Union's aims are the sole responsibility of the member states, which have entered into the treaty and are authorised to amend it again under Art. 48 TEU. However, Art. 3 TEU shall not be interpreted as the competence norm, and it does not permit the differentiation between the task/ aim norms on the one hand and competence norm, on the other hand, to be abandoned. Instead, to realise the Union's objectives and tasks, a decision-making competence in the TEU must be confirmed in each case, Art. 3 sub-para 6 TEU.[80] Among others, the following objectives should be mentioned.

The EU has taken the promotion of the people's well-being as one of its pilar objectives of the EU, Art. 3 (1) TEU.[81] This includes establishing an internal market based on balanced economic growth, which is difficult to envision without supporting vulnerable groups and mitigating the economic disparities between low-income households and those in better economic situations. A highly competitive market economy, which is also one of the objectives of the EU,[82] should aim high level of protection and improvement of the quality of the environment, Art. 3, (3) sub-para. 1 TEU.[83] This implies that the economy must be sustainable and environmentally friendly.[84]

The energy modernisation of the residential building stock is central to these objectives. Improved energy performance of dwellings contributes to higher living standards and thus, people's well-being. Additionally, the extensive renovation wave correlates with new job creation, positively impacting the economic situation. The consequential CO_2 emissions savings from such a wave significantly benefit the environment.

Moreover, the EU explicitly aims to combat social exclusion and discrimination, as stated in Art. 3 (3) sub-para. 2 TEU. Low-income households often face social exclusion and may experience direct or indirect discrimination, even at the legislative level. For example, Germany's approach to energy poverty in its National Energy and Climate Plan (NECP) is noteworthy. The German legislature does not address energy poverty, arguing that existing social and energy legislation adequately protects vulnerable households by guaranteeing them a minimum subsistence level, including covering their energy costs. As a result, the legislator fails to recognise the existing problem and thus does not address it through policy. With rising energy prices and the consequences of

[79] Pechstein in: Streinz 2018, Art. 3 EUV Rn. 2; Terhechte in: Grabitz/Hilf/Nettesheim 2021, Art. 3 EUV Rn. 15. Becker in: Becker et al. 2019, Art. 3 EUV Rn. 1f.

[80] Pechstein in: Streinz 2018, Art. 3 EUV Rn. 2; Becker in: Becker et al. 2019, Art. 3 EUV Rn. 1f.

[81] Terhechte in: Grabitz/Hilf/Nettesheim 2021, Art. 3 EUV Rn. 21f.

[82] Ibid, Rn. 56.

[83] Ibid, 2021Rn. 21f.

[84] Müller-Graff in: Pechsteint/Nowak/Häde 2017, Art. 3 EUV Rn. 39; Kahl et al. 2021, §2 Rn. 17.

the Covid-19 crisis, a significant social group may slide into the risk of social exclusion. Recognising the problem by the state would be a critical first step in addressing it.

Furthermore, the EU is mandated to integrate a high level of environmental protection and improve environmental quality in its policies, aligning with the principle of sustainable development. This follows from the Art. 3 (5) TEU and has become one of the pillars of the EGD and the "Fit for 55%" legislative package. This objective permeates into EU secondary law, where it is further defined and implemented. Accordingly, the Union's policies must incorporate a high level of environmental protection and environmental quality enhancement, underpinning the principle of sustainable development. Sustainable development, inherently linked to sustainable growth, cannot be achieved without addressing energy poverty and its impact on affected populations. As these objectives resonate through EU secondary law, they transform from mere ambitions to a new legal trajectory for the environmental reorientation of the EU legal order.[85] This sets the overarching framework for the secondary legislative acts that will be analysed subsequently.

Another crucial legal norm in the EU environment policy framework context is Art. 11 TFEU, known as the integration clause. According thereto, environmental protection requirements must be integrated into the definition and implementation of all Union policies and activities. This means that environmental and climate considerations should be interwoven with other policy areas, such as economic, transport, and agricultural policy, in accordance with sustainability principles.[86] Art. 11 TFEU can be interpreted as a primary law requirement for conducting strategic environmental impact assessments that encompass individual measures, policies, programmes, plans, and laws. This obligation extends to the legislative, executive, and judicial branches.[87] In light of Art. 11 TFEU, environmental protection requirements must be balanced against other concerns, such as economic freedom and welfare state considerations. The importance of Art. 11 TFEU is hardly overestimated when working on decarbonisation policies as they are a part of environmental protection policies. These policies must be developed in concert with other relevant areas, including economic and energy poverty mitigation strategies. The central focus of Art. 11 TFEU is on environmental protection, including a foundation of objectives and a prohibition against undermining them.[88] However, environmental protection should not be regarded as an absolute priority within the scope of Art. 11 TFEU. This legal norm does not provide a basis for prioritising environmental policy above all other policies.[89]

---

[85] Sikora 2021 p. 686.

[86] Calliess/Dross 2020, p. 457. The interesting fact is that the principle of sustainability takes its roots in German law, where it has been firstly developed in forestry law, see Winter in: Jans 2003, p. 21.

[87] Calliess/ Dross 2020, p. 461; Oppermann/Classen/Nettesheim 2016, §33 Rn. 19.

[88] Calliess/ Dross 2020, p. 461.

[89] Frenz 1997, p. 65; Kotzur in: Geiger/Khan/Kotzur, Art. 11 AEUV Rn. 7.

The requirements of environmental protection within the scope of Art. 11 TFEU results EU's objective to safeguard the environment, as outlined in Art. 191 TFEU, which will be characterised later. These requirements are enhanced by the EU and the member states' obligations derived from the fundamental rights, such as human life and health protection (Art. 3 CFR, 8 EHCR, Art. 2 (2) GG).[90] The necessity to protect these fundamental rights again becomes the engine power for the legislator to address energy poverty, particularly as vulnerable social groups affected often suffer from low energy efficiency in their dwellings and related health issues.[91]

4.2.2 Competencies

As mentioned earlier, the researched topic is multifaceted and relates to different areas of EU regulative policies. It is essential to distinguish between the EU competencies and the competencies of the member states when speaking about decarbonisation of the residential building sector. This distinction helps understand the appropriate level for action and, as noted, often frames national regulations. Thus, the clear distinguishing of the competences allows to fill the gaps at the national level that arise precisely because of the EU's lack of competence at this level.

Two policy branches of the EU are particularly relevant to the researched topic: environmental and energy efficiency policy. While the broader energy policy also influences the topic, its significance to this research is limited. The environment is a shared competence area between the EU and its member states, as stipulated in Art. 4 (2) lit. e) TFEU.[92] Environmental action is a global concern, necessitating policies that transcend national legislation.[93]

The TFEU confers a broad scope of legislative competencies on the EU in environmental action through Articles 191–193. These norms confirm and specify the EU's competencies in this area.[94] Bringing environmental action to the EU level helps to internalise external costs more effectively. Simultaneously, EU environmental action is limited by these regulations, which impose obligations on the member states.[95]A higher level of environmental protection encourages the member states even after the complete EU-wide harmonisation of this area, as stressed in the Art. 193 TFEU. However, the

---

[90] Calliess/ Dross 2020, p. 461.

[91] Walker/ Day 2012, p. 69f.; Boardman https://ec.europa.eu/clima/eu-action/european-green-deal/european-climate-law_en, 167f.

[92] Streinz in: Streinz 2018, Art. 4 AEUV, Rn. 8.

[93] Talus 2013, p. 175; Calliess in: Calliess/Ruffert 2016, Art. 191 Rn. 1.

[94] Calliess/ Dross 2020, p. 457; Woerdman/ Roggenkamp/ Holwerda 2015, p. 279f.; Saurer 2017, p. 1577; Talus 2013, p. 178.

[95] Talus 2013, p. 177.

Commission, other member states, and ultimately the European Court of Justice control any protectionist abuse by the member states.[96]

The EU is enabled to enact environmental regulations in all areas unless specific competence provisions of the treaty apply to a particular case. This does not imply that the EU has unlimited competence in the environmental sphere. Potential conflicts between the competence provisions of the treaties must be resolved according to the principle of limited authorisation.[97] One reason for this is that the competence provisions often provide different majority requirements for decision-making in the Council and different legislative procedures or rights of participation for other organs.[98] Other limitations of the EU environmental action can be found in the national sovereignty in matters of "tax and environment versus economic cost trade-offs reflected and formally enshrined in Art. 192 and 194 TFEU".[99]

Art. 191 (1) TFEU stipulates the main objectives that the EU's policy on the environment shall contribute to.[100] According to Art. 191 (2) TFEU, the EU policy on the environment shall aim at a high level of protection, considering the diversity of situations in the various regions of the Union. This gives the EU the competence to set different objectives, for instance, for decarbonising buildings, as it cannot be expected that economically weaker countries have the same capacities and bear the same burdens as the economically stronger ones.

In the context of building sector decarbonisation, the environmental aspects of the decarbonisation may conflict with the economic interests and property rights of the building owner, which are also protected by law. To balance these conflicting interests in respective regulations, two aspects must be considered. Additionally, the requirements of Art. 191 (1) and (2) TFEU, particularly the precautionary principle, must be taken into account. Moreover, the concept of integration does not permit the exclusion of environmental concerns. These requirements have gained importance as the environmental burden has increased. The legislator is not permitted to adopt measures that are recognisably designed to disproportionately disadvantage environmental protection, as this exceeds the scope of legislative freedom. A policy exceeding this limit would breach Art. 11 TFEU (which role for the environmental action will be analysed later) and, therefore, be unlawful.[101]

This implies that national legislators should regard environmental protection as a higher good in cases of conflicting interests when designing energy efficiency policies.

---

[96] Ibid, p. 178.

[97] Scherer/Haselhaus in: Dauses/Ludwig 2021, O. Umweltrecht, Rn. 74.

[98] Everling 1991, p. 181; Scherer/Haselhaus in: Dauses/Ludwig 2021, O. Umweltrecht, Rn. 74.

[99] Talus 2013, p. 175.

[100] Calliess/ Dross 2020, p. 457; Woerdman/ Roggenkamp/ Holwerda 2015, pp. 279–280; Talus 2013, p. 178f.; Kahl et al. 2021, §2 Rn. 9.

[101] Calliess/ Dross 2020, p. 461.

For this research, it means that the solution to the landlord-tenant dilemma must sufficiently consider environmental concerns, which may prevail over the interests of one of the parties. Otherwise, such a solution risks being inapplicable concerning the higher rank of EU law.

While Art. 191 TFEU sets out the objective of EU environmental policy and its principles, Art. 192 TFEU provides the corresponding competence provision in the sense of environmental action.[102] The term "action", as mentioned in Art. 192(1) TFEU, is a comprehensive term in the EU Treaties. It includes all legal acts within the meaning of Art. 288 TFEU, as well as other legal acts and measures without legal character (e.g., warnings, or communications from EU institutions).[103] Environmental protection is also mentioned in Art. 114 (3), (5) TFEU. The relation between Art. 114 (3) and the 192 (2) TFEU is, however, controversial.[104] This controversy, though, has no further substantive importance for the following research and will not be discussed in more detail.

Art. 192 (4) TFEU is particularly interesting for this research as it regulates the matter of financing and implementation of environmental policy. According to this legal norm, these competencies lie within the responsibility of the member states, unless the EU treaties specify otherwise. The EU's corresponding competence regarding the financing and implementation of environmental policy is included within this scope and can be inferred from the doctrine of implied powers.[105] The limitation of EU competence to "certain measures adopted by the Union" also reflects the general principle of subsidiarity. This implies that even if action at the EU level is more effective than at the national level, this should remain the exception in the case of financing measures. Conversely, the establishment of special financing instruments (e.g., the Cohesion Fund, Structural Fund, Just Transition Fund) at the EU level is permissible, as long as such financial measures have a general aim and are not "specific" within the meaning of Art. 192 (4) TFEU.[106]

As previously mentioned, the decarbonisation of the residential sector is closely interlinked with energy efficiency policy. The scope of regulation in this sphere is set in Art. 194 TFEU.[107] This article facilitates, among other aspects, the adoption of measures

---

[102] The earlier dispute about the appropriate competence provision within the environment chapter has been decided by the ECJ, see the Judgement of 11 June 1991, Titanium dioxide, C-300/89, ECLI:EU:C:1991:244; Scherer/Haselhaus in: Dauses/Ludwigs 2021, O. Umweltrecht, Rn. 75; Epiney 2019, §1 Rn. 1, §3 Rn.3.

[103] Nettesheim in: Grabitz/Hilf/Nettesheim 2021, Art. 175 AEUV, Rn. 11f.; Scherer/Haselhaus in: Dauses/Ludwigs 2021, O. Umweltrecht, Rn. 79.

[104] Schröder in: Streinz 2018, AEUV Art. 114, Rn. 139, 140; Korte in: Calliess/Ruffert 2016, AEUV Art. 114, Rn. 145; Rossi in: Vedder/Heintschel, Art. 114 AEUV, Rn. 2; Kahl in Streinz 2018, AEUV Art. 192, Rn. 95.

[105] Heselhaus 2001, p. 319f., 345f.; Scherer/Haselhaus in: Dauses/Ludwigs 2021, O. Umweltrecht, Rn. 100.

[106] Scherer/Haselhaus in: Dauses/Ludwigs 2021, O. Umweltrecht, Rn. 101.

[107] Granas 2013, p. 626;

under EU law to promote energy efficiency and savings, and to develop new and renewable energy sources. This area is also highly connected to environmental policy.[108] However, the EU's competence in the sphere of energy efficiency is confined to promoting energy efficiency and developing new and renewable forms of energy. The scope of this competence is limited insofar through the objectives mentioned in Art. 194 (1) TFEU, and these too are limited.[109] According to Art.194 (2) TFEU, without prejudice to the application of other provisions of the Treaties, the European Parliament and the Council are tasked with adopting the measures necessary to achieve these objectives. In other words, the member states retain considerable latitude to take actions in this sphere, while the EU provides only guidance for them.

The European energy efficiency policy has been subject to secondary legislation since the 1990s, based partly on internal market competence (now Art. 114 TFEU) and partly on environmental competence. Following the entry into force of the Treaty of Lisbon, the competence in the sphere of energy efficiency has fallen under the scope of Art. 194 TFEU, as its wording is relatively unambiguous.[110] Some German authors, however, continue to derive energy efficiency competence from environmental competence.[111] *Frenz* reasons this point of view as follows: considering that Art. 194(1) lit. c) TFEU refers to energy efficiency, energy saving, and associated alternative energy sources, its content focuses on energy technology aspects. According to his position this opinion clearly derives from the wording of the article. The objective of new and renewable energy sources development does not cover all the aspects of renewable energies but rather refers to the technological aspects.[112] The wording of Art. 194 (1) lit. c TFEU does not give an impulse for such a narrow interpretation of the EU competence in energy efficiency. Energy efficiency promotion has explicitly been mentioned in Art. 194 TFEU. Meanwhile, the objectives of Art. 191 TFEU (1)—"preserving, protecting and improving the quality of the environment", "combating climate change"—are formulated more broadly. They necessarily benefit from promoting energy efficiency but can also be achieved by other means. This indicates that Art. 194 TFEU is distinct from Art. 191 TFEU.[113] However, it should be noted that this dispute is largely theoretical in nature.

Nonetheless, it is not always possible to precisely determine the actual competence regarding measures related to the decarbonisation of the residential building stock, as they might fall under both competencies—environmental action and energy efficiency. The relevant case law pertaining to the delimitation of competencies is inconsistent and has not yet provided coherent rules, even though the contours of the delimitation

---

[108] Epiney 2019, §4 Rn. 8; Prall/ Ewer in: Koch/Hoffmann/Reese 2018, §9 Rn. 35f.

[109] Epiney 2019, §4 Rn. 8.

[110] Gundel in: Pechstein/Nowak/Häde 2017, AEUV Art. 194 Rn. 14; Martin 2014, p. 62.

[111] Frenz 2010, Rn. 4694; Frenz/Kane 2010, p. 470.

[112] Frenz/Kane 2010, p. 470.

[113] Caliess in: Caliess/Ruffert 2016, Art. 195 AEUV Rn. 14.

are becoming more evident in more recent case law.[114] There are some general rules for delimitation of the competencies outlined in the literature,[115] which, however, will not be addressed in detail in this research due to their lack of relevance.

It should be mentioned that Art. 194 (2) sub-para. 2 contains a far-reaching restriction of the comprehensive competence to act, which is anchored in the sub-para. 1. The scope of the Art. 194 (2) sub-para. 2 is broader than the similar provision of Art. 192 (2) lit. c TFEU. Firstly, the competence limit here is relevant not only when its conditions are cumulatively met. Secondly, Art. 194 (2) sub-para. 2 does not contain a materiality limit. Thirdly, Art. 194 (2) TFEU anchors a substantive competence limit, rather than merely being a procedural norm allowing unanimous decisions in the mentioned sphere, as is the case with Art. 192 (2) TFEU. In the case of Art. 194 (2) TFEU, the EU is entirely precluded from acting, even unanimously, for example, when a measure impacts the choice between different energy sources. In essence, the EU cannot formulate a comprehensive energy policy and must respect the preliminary decisions of the member states. The EU's energy policy framework can only be developed based on these preliminary decisions of the member states.[116]

Summarising the above, the EU sets the legal framework in the sphere of decarbonisation of residential building stock, which is closely linked to the climate mitigation and energy efficiency policies of the EU. However, the EU's competences in these spheres are quite limited, indicating that the decisions regarding the landlord-tenant dilemma in building stock decarbonisation are likely to be found at the national level. Nonetheless, this general framework of primary law is concretised by secondary law, which will be analysed later in Sub-chapter 4.3.

4.3 Secondary Law

4.3.1 Introduction and Ranking of Secondary Law Sources Related to the Decarbonisation of Dwellings Occupied by Low-income Tenants

As previously stated, EU primary law creates a general framework for energy efficiency policy, which is further developed through secondary law and, if necessary, transposed into the national law of member states if EU law does not have direct effect. As energy efficiency law is closely linked to environmental and climate law, it has evolved

[114] Kahl in Streinz 2018, AEUV Art. 192, Rn. 93.

[115] Kahl in Streinz 2018, AEUV Art. 192, Rn. 91; Calliess in: Calliess/Ruffert 2016, Art. 194 AEUV Rn. 3; Calliess in: Calliess/Ruffert 2016, Art. 192 AEUV Rn. 24f..; Kahl et al. 2021, §2 Rn. 38f.

[116] Calliess in: Calliess/Ruffert 2016, AEUV Art. 194, Rn. 29.

dynamically in recent years. The main sources of EU secondary law related to the decarbonisation of dwellings occupied by low-income tenants will be briefly analysed below. Depending on their character, they are either applied directly or oblige the member states to transpose them into their national legal orders. In principle, member states are also permitted to implement stricter measures than those prescribed at the EU level, provided EU law is not undermined. Understanding this scope allows us to assess whether the landlord-tenant dilemma is sufficiently addressed at this level and if countries have implemented adequate measures in their national orders. Regulatory policy recommendations can then be inspired based on the results of this research. However, those acts that only indirectly impact energy efficiency policy are beyond the scope of this thesis.

The main catalogue of secondary law sources relevant to this research topic is provided in Art. 288 TFEU and includes regulations, directives, decisions, recommendations, and opinions referred to therein.[117] These acts, often referred to as derived law ("*abgeleitetes Recht*"), are adopted by EU institutions based on the Treaties. However, secondary law sources also include atypical legal acts not mentioned in the Treaties, which the institutions may adopt in certain cases.[118] These should not be confused with soft law, i.e., Union acts not intended to have legal effects.

It is essential to clarify at this stage the ranking of the secondary law sources within the EU legal order to avoid later the question of whether any concrete legal regulation referred to the decarbonisation of the building stock prevails another one. As follows from Art. 263, 267 (1) lit. b), 288 (1) and 216 (2) TFEU, secondary law ranks below primary law and the international agreements of the EU, and they make the scope of the measure, therefore. The Treaties, particularly Art. 288 TFEU, do not stipulate a hierarchy among secondary legal acts. There are only some indications of such gradation in the case law of the ECJ, suggesting that basic legal acts should prescribe the essential elements of implementing legal acts.[119]

---

[117] Schroeder in: Streinz 2018, Art. 288 AEUV Rn. 29; Nettesheim in: Grabitz/Hilf/Nettesheim 2021, Art. 288 AEUV Rn. 209; Ruffert in: Calliess/Ruffert 2016, Art. 288 AEUV Rn. 98; Simon 2001, Rn. 271.

[118] The atypical legal acts are not mentioned in Art. 288 TFEU are often described as "cooperative Union law"; however, this term is also misleading as it does not exclusively refer to measures of cooperation between the Union institutions and the member states. These atypical legal acts include inter-institutional agreements, intra-institutional law, resolutions, communications and other administrative provisions and declarations, see more: Schroeder in: Streinz 2018, Art. 288 AEUV Rn. 29; Nettesheim in: Grabitz/Hilf/Nettesheim 2021, Art. 288 AEUV Rn. 209; Ruffert in: Calliess/Ruffert, Art. 288 AEUV Rn. 98; Simon 2001, Rn. 271.

[119] Court of Justice, Judgement of 22 January 2014, C-270/12, United Kingdom of Great Britain and Northern Ireland v. European Parliament and Council of the European Union, ECLI:EU:C:2014:18, para. 48; Ruffert in: Calliess/Ruffert 2016, Art. 288 AEUV Rn. 10.

The prevailing opinion in literature disputes the possibility of a norm hierarchical ranking.[120] According to this opinion, there is no reason to subordinate decisions to directives or regulations, for example. Nor is there any institutional hierarchy according to which legal acts of one EU institution (e.g., of the Commission) shall have a higher ranking than legal acts of another institution (e.g., of the Council).[121] The rule, according to which the *volonté générale* is expressed in the law, is prior to other legislative acts, which is not applicable to the EU legislation. Moreover, there is no democratic legitimation for such a ranking. Additionally, EU legislation cannot be equated with national parliamentary legislation, even with the intensive participation of the European Parliament.[122] Thus, any collision must be resolved according to general principles, where a *lex posterior* amends an older regulation, or a *lex specialis* amends a more general one.[123] However, legal acts serving to concretise secondary law rank below the provisions of the legal act from which they are derived. Implementing or delegated acts are derived from the competence to enact a basic legal act (cf. Art. 290, 291 TFEU). An attempt to establish a hierarchy of legal acts,[124] will not be addressed here.

The primary acts on energy efficiency on the EU level have been adopted in the form of directives, granting member states discretionary powers in their transposition into national law.[125] Meanwhile, implementing acts are usually enacted as regulations.[126] The following sub-chapter will briefly remind the reader about the scope of the direct effect of EU secondary law, focusing only on directives and regulations as they are the most common forms regulating the decarbonisation of the building stock. The general primacy of the application of EU law is also beyond the scope of this research.

According to Art. 288 (2) TFEU, regulations have general application. They are binding and directly applicable in all member states. This means that the regulations have a direct effect from the outset and do not require any participation of the member states. Therefore, neither a national transposition nor the national announcement of regulation is necessary (though often helpful).[127] Furthermore, regulations do not require

---

[120] Ruffert in: Calliess/Ruffert 2016, Art. 288 AEUV Rn. 10; Oppermann/Classen/Nettesheim 2016, §11 Rn. 46; Nettesheim in: Grabitz/Hilf/Nettesheim 2021, Art. 288 AEUV Rn. 227.

[121] Schroeder in: Streinz 2018, Art. 288 AEUV Rn. 24.

[122] Ruffert in: Calliess/Ruffert 2016, Art. 288 AEUV Rn. 10; Bast in: Bogdandy/Bast 2009, p. 654 f.

[123] Schroeder in: Streinz 2018, Art. 288 AEUV Rn. 24; Oppermann/Classen/Nettesheim 2016, §11 Rn. 46; Nettesheim in: Grabitz/Hilf/Nettesheim 2021, Art. 288 AEUV Rn. 228.

[124] The European Convention, Final Report of Working Group IX on Simplification, CONV 424/02 of 29.11.2002, p. 8f.; Schroeder in: Streinz 2018, Art. 288 AEUV Rn. 23; Biervert in: Becker et al. 2019, Art. 288 AEUV Rn. 21.

[125] Gundel in: Dauses/Ludwigs 2021, Energierecht Rn. 247; Härtel in: Härtel/ Pielow 2014, p. 32.

[126] Härtel in: Härtel/ Pielow 2014, p. 31f.

[127] ECJ, Judgement of 7 November 1972, Belgian State v. NV Cobelex, Case 20–72, ECLI:EU:C:1972:94; Nettesheim in: Grabitz/Hilf/Nettesheim 2021, Art. 288 AEUV Rn. 228;

a specific order of execution, as is the case in German law for the entry into force of international treaties, according to Art. 59(2) GG.[128] Due to the direct applicability, individuals can directly invoke a provision of a regulation under general conditions.[129] This does not apply if the concretisation of the provision is subject to an implementing regulation or if national authorities have been given a scope of discretion, as the EU legislator can authorise a member state to adopt implementing measures of a regulation.[130] Furthermore, national provisions or practices of the member states cannot be invoked against regulation. However, it is the responsibility of the member states to adopt appropriate national measures to ensure the full applicability of a regulation if necessary.[131]

Under Art. 288 (3) TFEU, directives are binding; however, each member state to which a directive is addressed has discretion concerning the choice of form and methods of its transposition into the national legal order. In this context, the principle of *effet utile* (practical effectiveness), anchored in Art. 4 (3) TEU, should also be mentioned. According thereto, member states must choose forms and methods of transposition that do not undermine the content of the respective directive. Over time, several judicial guidelines regarding the "how" of transposition have been formulated and must be considered by the member states. The transposition of the directives must meet requirements of completeness, legal certainty, legal clarity, and controllability.[132] According to the ECJ, the directives should be transposed by indisputably binding measures.[133]

Although directives must be transposed into the national legislation, their legal provisions might have a direct effect in the case of an implementation deficit.[134] As there have been no precedents in the sphere of environmental protection or energy efficiency where any provision of any directive had a direct effect on Germany or Sweden, the preconditions for such an effect will be beyond the scope of this research.

As emphasised above, there is no actual hierarchical rank among secondary law sources, so the chronology of acts listed and analysed in this chapter has been determined based on the following criteria: the regulations will be analysed first as they have a direct effect for the member states. The *leges generales* essential for this research will be analysed before the *leges speciales*.

---

128 Nettesheim in: Grabitz/Hilf/Nettesheim 2021, Art. 288 AEUV Rn. 101.

129 Biervert in: Becker et al. 2019, Art. 288 AEUV Rn. 21.

130 ECJ, Judgement of 11 November 1992, Roland Teulie c Cave cooperative "les Vignerons de Puissalicon", Case C-251/91, ECLI:EU:C:1992:430, para. 13; Biervert in: Becker et al. 2019, Art. 288 AEUV Rn. 21.

131 Biervert in: Becker et al. 2019, Art. 288 AEUV Rn. 21.

132 Kahl et al. 2021, §2 Rn. 65.

133 Judgement of 12 December 1996, Quality required of shellfish water, C-298/95, ECLI:EU:C:1996:501, Rn. 16; Kahl et al. 2021, §2 Rn. 65.

134 Schroeder in: Streinz 2018, Art. 288 AEUV Rn. 86.

4.3.2 The European Climate Law

One of the regulations enacted to realise the goals set in the EGD[135] is the Regulation (EU) 2021/1119 of the European Parliament and of the Council of 30 June 2021 establishing the framework for achieving climate neutrality and amending Regulations (EC) No 401/2009 and (EU) 2018/1999 ("European Climate law") (ECL). It is based on the environmental competence derived from Art. 192 (1) TFEU and directly affects the member states as a legal act in the form of a regulation.[136]

The ECL is essential for this research because it establishes a framework for the irreversible and gradual reduction of anthropogenic greenhouse gas emissions by sources and enhancement of removals by sinks regulated in the EU law (Art. 1 (1) ECL). It not only sets out a binding objective of climate neutrality in the EU by 2050 but also a binding Union target of a net domestic reduction in greenhouse gas emissions for 2030 (Art. 1 (2) ECL).[137] This target underscores the significance of building stock decarbonisation across the EU, particularly in cases where the decarbonisation measures exceed the economic scope of interest of any party.

The ECL stipulates several considerations for setting a trajectory towards achieving climate neutrality. The Commission shall consider, among other things, cost-effectiveness and economic efficiency, investment needs and opportunities, and just and socially fair transition, as it is set in Art. 3 (3).[138] It also sends a clear message to member states to take actions in this direction. The cap of CO_2 equivalent to 225 million tonnes within the contribution of net removals to the Union set up in Art. 1 (1) sub-para 3 ECL, further emphasises the importance of decarbonising the building stock.

Under Art. 4 (2) sub-para. 1, the ECL obligates the Commission to review relevant EU legislation to achieve the intermediate and long-term EU climate targets. uring this review, and in future reviews, the Commission shall assess the availability of adequate instruments and incentives under EU law to mobilise the necessary investment and propose measures as required (Art. 4 (2) sub-para. 2 ECL). New financial instruments and schemes aiming to mobilise both public and private investment into the decarbonisation of the building sector are now of high importance. The instruments proposed later for Germany might be transposed in other legal orders. Further, this thesis could contribute to the achievement of the objectives of the ECL under Art. 7 (1), as it defines and offers solutions for the problems of the German regulatory status-quo related to the landlord-tenant dilemma in the context of vulnerable social groups, which inhibits building stock decarbonisation.

---

[135] 'European Green Deal'. n.d. Text. European Commission - European Commission. Accessed 14 November 2021. https://ec.europa.eu/clima/eu-action/european-green-deal/european-climate-law_en

[136] Schlacke/Köster/Thierjung 2021a, p. 621; Frenz 2020, p. 614.

[137] Frenz 2020, p. 614.

[138] COM (2020) 80 final; Falke 2021, p. 440.

Overall, while the ECL has a formal and procedural nature, it also amends Regulations (EC) No 401/2009[139] and (EU) 2018/1999.[140] Apart from establishing the binding goal of climate neutrality, it does not contain further concrete goals. However, it establishes a mechanism through which the Union can strongly influence the member states and acquire the instruments to efficiently achieve the goal of greenhouse gas neutrality by 2050 (Art. 7 ECL).[141] For the researched topic, the ECL's binding climate objectives are of the highest relevance as they oblige the member states to take decarbonisation actions. At this point, the following thesis can contribute to this task inspiring new regulative policies in the researched sphere.

4.3.3 The EU Taxonomy Regulation

The subsequent relevant regulation for the researched context is Regulation (EU) 2020/852 of the European Parliament and of the Council of 18 June 2020 on the establishment of a framework to facilitate sustainable investment and amending Regulation (EU) 2019/2088 (the EU Taxonomy). The OECD has estimated that approximately $6.9 trillion (about €6.35 trillion) in annual investment is required globally to achieve the Paris Agreement goals by 2030.[142] The role of finance in energy transition and decarbonisation is hard to overestimate. The EU Taxonomy's primary objective is to provide member states with a classification system for sustainable activities,[143] contributing to increased sustainable finance, such as the Sustainable Finance Strategy and the European Green Bond Standard.[144] One effective instrument for financing the decarbonisation of the building sector, as evidenced by other countries' experiences, is the issuance of green bonds. This will be examined and discussed later in Chap. 7. In this context, the EU Taxonomy contributes by providing member states, including Germany, with a classification of green instruments, which enhances the attractiveness of using green bonds.

---

[139] Regulation (EC) No 401/2009 of the European Parliament and of the Council of 23 April 2009 on the European Environment Agency and the European Environment Information and Observation Network.

[140] Regulation (EU) 2018/1999 of the European Parliament and of the Council of 11 December 2018 on the Governance of the Energy Union and Climate Action, amending Regulations (EC) No 663/2009 and (EC) No 715/2009 of the European Parliament and of the Council, Directives 94/22/EC, 98/70/EC, 2009/31/EC, 2009/73/EC, 2010/31/EU, 2012/27/EU and 2013/30/EU of the European Parliament and of the Council. Council Directives 2009/119/EC and (EU) 2015/652 and repealing Regulation (EU) No 525/2013 of the European Parliament and of the Council.

[141] Frenz 2020, p. 617.

[142] OECD 2017, p. 102; EU Technical Expert Group on Sustainable Finance 2020, p. 7.

[143] EU Technical Expert Group on Sustainable Finance 2020, p. 7.

[144] 'Sustainable Finance'. n.d. Text. European Commission - European Commission. Accessed 2 December 2023. https://ec.europa.eu/commission/presscorner/detail/en/ip_21_3405.

The EU Taxonomy particularly establishes criteria for determining whether an economic activity is environmentally sustainable, thereby establishing how environmentally sustainable an investment in that activity is, as per Art. 1 (1). This regulation adopted as part of the EGD, is one of the legal documents adopted due to the EGD, implements the Action Plan on Financing Sustainable Growth published by the European Commission in 2018.[145]

One of the proposed solutions for addressing the landlord-tenant dilemma involves mobilising third-party financial assets, for example, through issuing green bonds. It is crucial to define what constitutes these green bonds and which types of energy modernisation financed through this instrument are sustainable. The EU Taxonomy Regulation is significant in this regard, especially since it has a direct effect and does not require incorporation into national legislation. Economic activities related to the decarbonisation of the building stock can be assessed according to Art. 10 of the EU Taxonomy to determine if they contribute substantially to climate change mitigation by improving energy efficiency. Art. 11 (1) defines when an economic activity significantly contributes to climate change adaptation. The technical annex to the EU Taxonomy Regulation provides technical screening criteria for building renovation and substantial contribution to climate change mitigation and adaptation.[146]

However, the EU Taxonomy Regulation has been criticised in literature for being ineffective.[147] The main categories of the EU Taxonomy Regulation are the categories of "sustainable" and "non-sustainable" activities, which are expected to be included in future extended non-financial or consolidated non-financial statement information of companies. This reporting will detail how and to what extent the undertaking's activities are associated with environmentally sustainable economic activities.[148]

A central issue is the concept of "economic activity", encompassing both financial and non-financial undertakings as per Art. 8 (1) of the EU Taxonomy Regulation. Art. 8 aims to increase market transparency for green finance and prevent greenwashing by providing information about the environmental performance of assets and economic activities.[149] However, the regulation does not precisely define "economic activity",[150] potentially opening the door for greenwashing, which the EU Taxonomy Regulation

---

[145] Raux/Fischer 2021, p. 16.

[146] EU Technical Expert Group On Sustainable Finance. 2020. 'Taxonomy: Final Report'. https://finance.ec.europa.eu/system/files/2020-03/200309-sustainable-finance-teg-final-report-taxonomy_en.pdf.

[147] Ekkenga/Schirmacher/Schneider 2021, p. 1511; Lamy/Bach 2020, p. 349.

[148] Ibid.

[149] European Commission. n. d. 'FAQ. What Is the EU Taxonomy Article 8 Delegated Act and How Will It Work in Practice?' https://ec.europa.eu/info/sites/default/files/business_economy_euro/banking_and_finance/documents/sustainable-finance-taxonomy-article-8-faq_en.pdf., p. 3.

[150] Ekkenga/Schirmacher/Schneider 2021, p. 1511; Lamy/Bach 2020, p. 349.

aims to tackle.[151] Art. 16 only mentions that an economic activity shall qualify as contributing sustainability to one or more environmental objectives set out in Art. 9 by directly enabling other activities to make a sustainable contribution to one or more of those objectives. Such an economic activity shall not lead to a lock-in of assets that undermine long-term environmental goals, considering the economic lifetime of those assets, and it shall have a substantially positive impact based on life-cycle considerations. This description does not provide yet any concrete points. Economic activity is also mentioned in recitals, for example, in recital No. 43 of the EU Taxonomy Regulation. According to it, the Commission should consider the statistical classifications relating to the environmental goods and services sector, namely the classification of environmental protection activities (CEPA) and the classification of resource management activities (CRMA) of Regulation (EU) No 538/2014 of the European Parliament and of the Council to avoid unnecessary inconsistencies with classifications of economic activities. The Court of Justice of the European Union, however, has commented on the legal quality of recitals of the EU legal acts, particularly in its decision of 19 June 2014 in Case C-345/13 that recitals have no legally binding effect.[152]

Furthermore, obligations under Art. 8 apply primarily to large financial institutions with an average number of 500 employees, as outlined in Art. 8 (1) of the EU Taxonomy in conjunction with Art. 19a or 29a of Directive 2013/34/EU. This issue remains crucial for the issuers of green bonds, as large banks are often more active in using innovative instruments than smaller banks. Various strategies can be pursued with different interpretations of “economic activity”. For instance, companies might engage in “cherry-picking”, highlighting only the most sustainable aspects of their business in corporate social responsibility (CSR) reports.[153] As an example, the producer of construction materials whose production share of sustainable materials accounts only for 10% of total production highlights precisely this share in the CSR report.[154] This strategy can also be used conversely: the term economic activity can be interpreted very broadly: the same producer can highlight as many parts of one’s performance programme as ecologically and sustainable as possible and do not mention the share of non-sustainable activity in the report.[155]

Defining “economic activity” in subsequent amendments is vital to achieve climate and energy goals, particularly in the building sector, and transition the economy towards sustainability. Nonetheless, the EU Taxonomy Regulation establishes a regulatory framework for financial instruments, particularly green bonds, that apply to the

[151] Ekkenga/Schirmacher/Schneider 2021, p. 1511.

[152] Deutscher Bundestag, PE 6 – 3000 – 020/19 (12.2.2019), p. 2.

[153] Ekkenga/Schirmacher/Schneider 2021, p. 1511.

[154] The example is based on Ekkenga/Schirmacher/Schneider 2021, p. 1511.

[155] Ekkenga/Schirmacher/Schneider 2021, p. 1511.

decarbonisation of dwellings. This will be explored in Chap. 7 of this thesis. Other critical issues related to the EU Taxonomy Regulation are beyond the scope of this thesis.

4.3.4 The Effort Sharing Regulation

The Regulation (EU) 2018/842 of the European Parliament and of the Council of 30 May 2018 on binding annual greenhouse gas emission reductions by the Member States from 2021 to 2030 contributing to climate action to meet commitments under the Paris Agreement and amending Regulation (EU No 525/2013) (the Effort Sharing Regulation) is one of the critical pillars of the climate mitigation regulative framework on the EU level.[156] It is intended to serve the implementation of the Energy and Climate Package, thereby serving as a central instrument for enacting the Paris Agreement within the European Union.[157] The ESR compels member states to meet their commitments to achieving greenhouse gas emission reductions, including those from the energy performance of buildings as highlighted in recital 8 of the ESR. The significance of the ESR for this research lies in its establishment of national targets for emission reduction, particularly from the heating of buildings, as set out in Art. 2 (1) ESR.[158]

According to Art. 4 (1) ESR, each member state is required, by 2030, to limit its overall greenhouse gas emissions by at least the percentage specified in Annex I, relative to its greenhouse gas emissions in 2005. For Germany, this target is set at a reduction of 38%, and for Sweden, it is 40%. Consequently, the compared countries, Germany and Sweden, have similar objectives. This similarity raises the expectation that these countries are at a comparable level of decarbonisation of their building stock, especially concerning the amount of CO_2 emissions produced in this sector. Should one country demonstrate best practices in addressing the landlord-tenant dilemma, there is a heightened likelihood that such experiences could be positively transposed into another legal order, particularly when they share a similar context.

4.3.5 The Regulation on the Energy Union Governance

The Regulation on the Governance of the Energy Union and Climate Action, amending Regulations (EC) No 663/2009 and (EC) No 715/2009 of the European Parliament and the Council, Directives 94/22/EC, 98/70/EC, 2009/31/EC, 2009/73/EC, 2010/31/EU, 2012/27/EU and 2013/30/EU of the European Parliament and the Council, Council

[156] Peeters/ Athanasiadou 2020, p. 202.

[157] Ekardt et al. 2018, p. 149.

[158] 'Effort Sharing Regulation'. n. d. Text. European Commission – European Commission. Accessed 22 December 2023. https://ec.europa.eu/commission/presscorner/detail/en/qanda_21_3543.

Directives 2009/119/EC and (EU) 2015/652 and repealing Regulation (EU) No 525/2013 of the European Parliament and the Council (the Governance Regulation) establishes a governance mechanism for implementing strategies and measures to meet energy and climate objectives and targets consistent with the Paris Agreement and other relevant legislative acts (Art. 1(1) sub-para. 1 lit. a) of the Governance Regulation). This regulation is a part of the comprehensive legislative package "Clean Energy for all Europeans", designed to update the energy policy framework to facilitate the transition to clean energy and deliver on the EU's Paris Agreement commitments for greenhouse gas emission reduction.[159] The Governance Regulation acts as an "umbrella legal act" for comprehensive governance of energy and climate policy from 2021 to 2030, primarily aiming to implement the climate protection goals of the Paris Agreement and to compensate for the lack of EU competencies in energy supply.[160] The Governance Regulation has successfully linked EU energy and climate policy, which was necessary to do in the light of the EU's obligation under international law to meet the climate protection goals of the Paris Agreement.[161]

For the researched topic, the Governance regulation has a high relevance for several reasons. Firstly, it applies to dimensions of the Energy Union such as energy efficiency and decarbonisation (Art. 1 (2) lit. c), lit. d), providing directly applicable law relevant to these spheres and contributing to the general framework of energy efficiency and decarbonisation. Secondly, the governance mechanism is based on long-term strategies, integrated national energy and climate plans covering ten-year periods starting from 2021 to 2030, corresponding integrated national energy and climate progress reports by member states, and integrated monitoring agreements by the Commission (Art. 1 (1) sub-para. 2 of the Governance Regulation). This ensures that national actions of member states are considered at the EU level, increasing the importance of academic discussions related to energy poverty in Germany, especially in relation to the landlord-tenant dilemma. Notably, Germany has yet to address energy poverty in its integrated national energy and climate plans, potentially hindering legislative and executive responses to this issue. However, the regulation specifies in Art. 3 (2) and Annex I the elements that the national energy and climate plans should contain, including the number of households in energy poverty, necessary domestic energy services to guarantee basic living standards, and existing social and other relevant policies, with indicative Commission guidance on relevant indicators for energy poverty.[162]

---

[159] Knodt/ Schlacke 2019, p. 404.

[160] Ibid, p. 404 f.

[161] Ibid, p. 408.

[162] However, due to the insufficient authorisation through primary law to design energy policy, the Governance Regulation uses soft mechanisms to coordinate the implementation of energy and climate policy objectives at the member state level. This makes the Governance Regulation strongly dependent on the will of the member states, as the regulation does not specify how the measures,

Furthermore, the Governance Regulation anchors the main objectives, targets and contributions to be included in the national energy and climate plans for each dimension, as detailed in Art. 4. Regarding the "Energy Efficiency" dimension, member states are required to submit, among other aspects, the cumulative amount of end-use energy savings to be achieved over the period 2021–2030; indicative milestones of the long-term strategy for renovating the national stock of residential and non-residential buildings, both public and private; a roadmap with measurable progress indicators; an evidence-based estimate of expected energy savings and broader benefits; and contributions to the Union's energy efficiency targets according to Directive 2012/27/EU. They also need to report on the total floor area to be renovated or equivalent annual energy savings to be achieved from 2021 to 2030 under Art. 5 of Directive 2012/27/EU, focusing on the exemplary role of public bodies' buildings. These requirements underscore the relevance of the researched topic and the importance of efficient solutions for decarbonising the residential building stock, thereby creating a framework for the implementation of targeted measures.

4.3.6 The Energy Performance of Buildings Directive

As previously mentioned, directives generally must be transposed into the national legislation of the member states to provide binding effects for individuals. Member states usually have discretionary powers regarding the transposition of these legal acts and may even implement stricter measures than those prescribed by the directives themselves. However, directives create a framework for national legislators,[163] helping to harmonise legislation in respective spheres across member states. European energy efficiency law is characterised by its sectoral design, with legal norms regulating the building sector primarily focusing on thermal insulation, efficient heating and hot water systems, and ventilation systems. These norms establish only the minimum requirements for the primary energy consumption of new buildings and existing ones undergoing significant renovations.[164]

Two directives are of special importance for the energy efficiency of the building stock, namely the Energy Performance of Buildings Directive and the Energy Efficiency Directive. They both are part of the energy efficiency legislation of the EU and pursue

for example, the integrated national energy and climate plans have to be implemented, see Knodt/ Schlacke 2019, p. 408.

[163] For Germany, it formed in the past he legal framework for the EnEG and the EnEV and now for the GEG, see Topp in Theobald/Kühling 2021, Energierecht, Gebäuderichtlinie, Rn. 77.

[164] Härtel in: Härtel/ Pielow 2014, p. 35.

the same objectives. They both are interlinked to each other,[165] providing essential definitions for national legislators.

Directive 2010/31/EU of the European Parliament and of the Council of 19 May 2010 on the energy performance of buildings (the Energy Performance of Buildings Directive, the EPBD) seves as the *lex specialis* in matters of energy efficiency in the building sector.[166] This is, however, not the first legal act on the EU level to set a framework for the energy efficiency of buildings. The first EU directive to set a framework for building energy efficiency was the Energy Performance of Buildings Directive 2002/91/EC, introducing the "whole building" approach. Member states were required to adopt a national or regional methodology for calculating building energy performance based on the EU framework and apply minimum energy performance requirements to new and majorly renovated large buildings. The recast of the EPBD in 2009 introduced a new cost optimality concept in energy performance.[167]

As buildings account for 40% of total EU energy consumption[168] and the building sector has been expanding in recent years, it is expected to continue growing in the coming years to meet the pressing demand in the housing market. This implies that an increase in energy consumption is anticipated unless changes occur. The entire legislation of the European Union is being updated in line with the latest science and technology to meet the EU's climate and energy objectives. The role of the EPBD in this research is not only to provide the legal framework for the assessed topic but also to identify paths for potential solutions that Germany has not yet implemented, but which can already contribute to designing possible solutions.

One of the guiding principles of the EPBD is the "energy efficiency first" principle, which plays a significant role in achieving energy and climate goals. Directive 2010/31/EU already emphasised the importance of energy efficiency improvement in the building sector over a decade ago. In 2018, this directive was amended by Directive (EU) 2018/844 of the European Parliament and of the Council of 30 May 2018, amending Directive 2010/31/EU on the energy performance of buildings and Directive 2012/27/EU on energy efficiency. Better-performing buildings provide higher comfort levels and well-being for their occupants and improve health, as proven by the 2009 World Health Organisation guidelines, recital No. 13. The amended directive aims to promote financial mechanisms, incentives, and the mobilisation of financial institutions for energy efficiency renovations in buildings, according to recital No. 16. The forthcoming research can contribute to these aims as it proposes a solution to the landlord-tenant dilemma, addressing the vulnerable group of people suffering from energy poverty.

[165] Topp in Theobald/Kühling 2021, Energierecht, Gebäuderichtlinie, Rn. 77.

[166] Topp in Theobald/Kühling 2021, Energierecht, Gebäuderichtlinie, Rn. 77.

[167] Economidou et al. 2011, p. 76.

[168] But also of the German consumption, see Koch 2011, p. 644.

The framework for the major renovation of existing buildings set up in Art. 7 of the EPBD is also important for this research, as it introduces the minimum energy performance requirements established under Art. 4, insofar as this is technically, functionally, and economically feasible. The directive itself does not define when the minimum energy performance is not feasible. However, Annex III of the EPBD introduces a comparative methodology framework for identifying cost-optimal levels of energy performance requirements for buildings and building elements. These requirements have been transposed into German legislation and will be analysed in the context of the landlord-tenant dilemma in Chap. 5.

The EPBD is one of the important pieces of EU legislation creating a framework for national legislation in drafting renovation strategies and addressing specific issues such as landlord-tenant-dilemma and energy poverty. This increases the benefits of energy renovation. However, it remains challenging for member states to design their strategies as ambitiously as possible while simultaneously reducing barriers for implementers at the local level,[169] as the respective analyses of German legislation will demonstrate in Chap. 5.

4.3.7 The Energy Efficiency Directive

Lastly, the Directive 2012/21/EU of the European Parliament and of the Council on energy efficiency, amending Directives 2009/125/EC and 2010/30 EU (The Energy Efficiency Directive, the EED), established a set of binding measures to assist the EU in reaching its 20% energy efficiency target by 2020.[170] The EED was further amended by Directive (EU) 2018/2002 of the European Parliament and of the Council. The revised EED aims to reduce primary energy within the European Union by at least 32.5% by 2030, as per Art. 1 (1) EED. Additionally, the final energy savings obligation was extended and strengthened for the period beyond 2020. For the first time, new savings each year from 2021 to 2030 of 0.8% of annual final energy consumption in each member state were agreed upon, as stated in Art. 7 (1) EED.[171] As previously mentioned, the EED is closely linked to the EPBD, but in the context of the energy performance of buildings, this directive serves as *lex generalis*. The legal basis of the EED is Art. 194 (2) TFEU, which is the competence norm for energy measures.[172] The EED was further amended by Directive (EU) 2018/2002 of the European Parliament and of the Council. The revised EED aims to reduce primary energy within the European Union by at least 32.5% by 2030, as per Art. 1 (1) EED. Additionally, the final energy savings obligation

[169] Bean et al. 2019, p. 5.

[170] Scholtka/ Keller-Herder 2019, p. 897.

[171] Ibid.

[172] COM (2021) 558 final.

was extended and strengthened for the period beyond 2020. For the first time, new savings each year from 2021 to 2030 of 0.8% of annual final energy consumption in each member state were agreed upon, as stated in Art. 7 (1) EED.

Furthermore, the EED obliges member states to establish long-term strategies for mobilising investment to renovate the building stock. Among other requirements, these strategies should include a forward-looking perspective to guide investment decisions of individuals, the construction industry, and financial institutions, as per Art. 4 (1) EED. Member states are also required to evaluate and remove regulatory and non-regulatory barriers to energy efficiency. If a split incentive between landlords and tenants or among owners is identified, this barrier must be addressed, as stated in Art. 19 (1) EED. These obligations establish the framework for regulatory policies related to this research. Considering the social state principle of Art. 20 (1) GG, Germany faces a balancing act between addressing the landlord-tenant dilemma and the social affordability of energy modernisation costs for tenants. For low-income households, bearing the costs of energy modernisation is particularly challenging. Therefore, the landlord-tenant dilemma should be addressed in such a way that they are relieved of this burden. An innovative funding scheme could contribute to this relief.

Art. 9b (3) and Art. 10a (1) EED are also of essence for the proposed solutions as they oblige member states to allocate costs of heating in multi-apartment buildings in accordance with the actual individual consumption. This obligation shall be considered later when evaluating possible landlord-tenant dilemma solutions. For instance, the warm rent model could contradict with this requirement.

The European Green Deal has announced plans to revise relevant legislative measures to meet increased climate ambitions, including the EED by June 2021. In July 2021, a proposal for a new directive on energy efficiency was introduced by the Commission. The new EED will primarily evaluate elements not revised in 2018 and secondly, consider the impact assessment to achieve the increased 2030 greenhouse gas emission targets. It will assess whether the EED is suitable for overcoming remaining regulatory and non-regulatory barriers and market failures. Identified barriers, gaps, or weaknesses will be addressed through the new amendments.[173] It is expected that the EED will be reviewed in the nearest future; the process has been started.[174] The revision of the EED is expected to support energy efficiency and building renovation investment through the EU funding, which will help address energy poverty.[175]

[173] Council of European Energy Regulators 2021, p. 5.

[174] 'Commission Proposes New Energy Efficiency Directive - European Commission'. n.d. Accessed 22 December 2023. https://commission.europa.eu/news/commission-proposes-new-energy-efficiency-directive-2021-07-14_en.

[175] COM (2021) 558 final.

4.4 Conclusion

The EU regulatory framework in the sphere of energy efficiency and decarbonisation of dwellings is highly complex and intricately linked to overarching climate mitigation policies. It has undergone several modifications in the past decade to align with the ambitious climate protection and energy goals of the European Union. In light of the Russian invasion of Ukraine and the subsequent energy crisis in Europe, the importance of energy savings cannot be overstated. Therefore, the EU regulative serves as a lighting tower for member states on their journey towards climate and energy goals, as well as energy independence from Russia.

Primary law establishes the fundamental principles and the distribution of competences in the assessed areas. Basic regulatory policies are typically enacted within secondary law, primarily as directives that offer member states discretionary powers regarding their transposition. Implementing acts are often passed in the form of regulations and typically do not require transposition into national legislation. Due to the EU regulatory framework, legislation in member states in the researched field is harmonised to a certain extent. Nevertheless, they retain considerable leeway in deciding the measures for transposition and are encouraged to pursue stricter CO_2 reduction goals and apply more stringent standards, provided they do not contravene EU law. Given the complexity of the matter and the limited extent of EU competences in this area, communications as a form of soft law are often utilised, guiding member states towards the preferred policy direction.

Despite this framework, Germany, for example, does not recognise the existence of energy poverty, arguing that energy costs are covered by state social law regulations. Consequently, a significant portion of the affected population is overlooked in the design of national regulatory policies. This poses a risk that the measures developed by the member state may not sufficiently resolve the landlord-tenant dilemma.

Simultaneously, considering the comparative method of law, the experiences of member states, particularly those with similar contexts, can be highly valuable in addressing the landlord-tenant dilemma in Germany. The existing EU legal framework enables the identification of national margins for manoeuvre in addressing the issue and defines the scope of assessment in foreign legal systems. It also provides legal practitioners dealing with a foreign legal order with support when searching for solutions.

References

Appel, Ivo, and Georg Buchholz. 18. Handbuch Umweltrecht. Edited by Hans-Joachim Koch, Ekkehard Hofmann, and Moritz Reese. 5., Überarbeitete Auflage. München: C.H. Beck.

Bakardjieva Engelbrekt, Antonina. 2015. 'Comparative Law and European Law: The End of an Era, a New Beginning, or Time to Face the Methodological Challenges?' In , 87–104.

Bean, Frances, Camelia Rata, Sibyl Steuwer, Dragomir Tzanev, and Oliver Rapf. 2019. 'EU Energy Performance of Buildings Directive – Guidance for Public Officers. Navigating New

Requirements for Renovation Strategies'. https://bpie.eu/wp-content/uploads/2019/03/Our-Buildings_-EUKI_EPBD-Guidance_Final.pdf.

Becker, Ulrich, Armin Hatje, Johann Schoo, and Jürgen Schwarze. n. d. *EU-Kommentar*. 4th ed. Baden-Baden: Nomos. Accessed 24 November 2023. https://www.soldan.de/eu-kommentar-8067529.html.

Belgischer Staat gegen NV Cobelex. 1972. EuGH.

Bingener, Reinhard. 2022. 'Erdgas aus Russland: Wie Deutschland abhängig wurde'. *FAZ.NET*, 13 May 2022. https://www.faz.net/aktuell/politik/inland/erdgas-aus-russland-wie-deutschland-abhaengig-wurde-18025821.html.

Biondi, Andrea, Eglė Dagilytė, Anglia Ruskin, and Esin Küçük. 2018. *Solidarity in EU Law*. Edwar Elgar Publishing. https://www.e-elgar.com/shop/gbp/solidarity-in-eu-law-9781783477777.html.

Boardman, Brenda. 2010. Fixing Fuel Poverty: Challenges and Solutions. London. https://www.routledge.com/Fixing-Fuel-Poverty-Challenges-and-Solutions/Boardman/p/book/9781844077441.

Bogdandy, Armin, and Jürgen Bast, eds. 2009. *Europäisches Verfassungsrecht: Theoretische und dogmatische Grundzüge*. Springer-Lehrbuch. Berlin, Heidelberg: Springer. https://doi.org/10.1007/978-3-540-73810-7.

Calliess, Christian, and Miriam Dross. 2020. 'Umwelt- Und Klimaschutz Als Integraler Bestandteil Der Wirtschaftspolitik Überlegungen Im Lichte von European Green Deal Und Corona-Recovery Plan'. *ZUR*, no. 9: 456–64.

Calliess, Christian, and Matthias Ruffert. 2016. *EUV/AEUV. Das Verfassungsrecht Der Europäischen Union Mit Europäischer Grundrechtecharta*. 5th ed. C. H. Beck. https://beck-online.beck.de/?vpath=bibdata/komm/CalliesRuffertEUVEGV_4/cont/CalliesRuffertEUVEGV.htm.

'Commission Proposes New Energy Efficiency Directive – European Commission'. n. d. Accessed 22 December 2023. https://commission.europa.eu/news/commission-proposes-new-energy-efficiency-directive-2021-07-14_en.

'Commission Staff Working Document Analysis of the National Long-Term Renovation Strategies'. 2022. Brussels: European Commission. https://energy.ec.europa.eu/system/files/2022-12/SWD-Analysis-of-2020-LTRS.PDF.

Commission of the European Communities v Federal Republic of Germany. Failure by a Member State to fulfil obligations - Failure to transpose Directives 78/659/EEC and 79/923/EEC within the periods prescribed - Quality of fresh waters needing protection or improvement in order to support fish life - Quality required of shellfish waters. 1996. Europski sud.

'Commission Staff Working Document Preliminary Analysis of the Long-Term Renovation Strategies of 13 Member States'. 2021. Brussels. http://cdn.mitma.gob.es/portal-web-drupal/planes_estartegicos/commissionstaff.pdf.

Communication from the Commission to the European Parliament, the European Council, the Council, the European Economic and Social Committee and the Committee of the Regions Tackling Rising Energy Prices: A Toolbox for Action and Support. 2021. https://eur-lex.europa.eu/legal-content/EN/TXT/?uri=COM%3A2021%3A660%3AFIN.

'Commission Proposes New Energy Efficiency Directive - European Commission'. n. d. Accessed 22 December 2023. https://commission.europa.eu/news/commission-proposes-new-energy-efficiency-directive-2021-07-14_en.

Communication from the Commission to the European Parliament, the European Council, the Council, the European Central Bank, the European Economic and Social Committee, the Committee of the Regions and the European Investment Bank Annual Sustainable Growth Strategy 2021. 2020. https://eur-lex.europa.eu/legal-content/en/TXT/?uri=CELEX%3A52020DC0575.

Communication from the Commission to the European Parliament, the Council, the European Economic and Social Committee and the Committee of the Regions a New Circular Economy Action Plan for a Cleaner and More Competitive Europe. 2020a. https://eur-lex.europa.eu/legal-content/EN/TXT/?uri=COM%3A2020%3A98%3AFIN.

Communication from the Commission to the European Parliament, the Council, the European Economic and Social Committee and the Committee of the Regions a Renovation Wave for Europe – Greening Our Buildings, Creating Jobs, Improving Lives. 2020b. https://eur-lex.europa.eu/legal-content/EN/TXT/?uri=CELEX%3A52020DC0662.

Communication from the Commission to the European Parliament, the Council, the European Economic and Social Committee and the Committee of the Regions an EU Strategy on Heating and Cooling. 2016. https://eur-lex.europa.eu/legal-content/EN/TXT/?uri=CELEX%3A52016DC0051.

Communication from the Commission to the European Parliament, the Council, the European Economic and Social Committee and the Committee of the Regions 'Fit for 55': Delivering the Eu's 2030 Climate Target on the Way to Climate Neutrality. 2021a. https://eur-lex.europa.eu/legal-content/EN/ALL/?uri=COM%3A2021%3A550%3AFIN.

Communication from the Commission to the European Parliament, the Council, the European Economic and Social Committee and the Committee of the Regions 'Fit for 55': Delivering the Eu's 2030 Climate Target on the Way to Climate Neutrality. 2021b. https://eur-lex.europa.eu/legal-content/EN/ALL/?uri=COM%3A2021%3A550%3AFIN.

Communication from the Commission to the European Parliament, the Council, the European Economic and Social Committee and the Committee of the Regions Stepping up Europe's 2030 Climate Ambition Investing in a Climate-Neutral Future for the Benefit of Our People. 2020. https://eur-lex.europa.eu/legal-content/EN/TXT/?uri=CELEX%3A52020DC0562.

Communication from the Commission to the European Parliament, the European Council, the Council, the European Economic and Social Committee and the Committee of the Regions Repowereu Plan. 2022. https://eur-lex.europa.eu/legal-content/EN/TXT/?uri=COM%3A2022%3A230%3AFIN.

Communication from the Commission to the European Parliament, the European Council, the Council, the European Economic and Social Committee and the Committee of the Regions the European Green Deal. 2019. https://eur-lex.europa.eu/legal-content/EN/TXT/?uri=COM%3A2019%3A640%3AFIN.

Council of European Energy Regulators. 2021. 'Public Consultation on the EU Energy Efficiency Directive (EED). CEER Response for the European Commission'. Ref: C21-CRM-141–05. https://www.ceer.eu/documents/104400/-/-/d52f8d36-0d08-27b2-eef9-52f66520619c.

Dauses, Manfred, and Markus Ludwig. 2021. *Handbuch des EU-Wirtschaftsrechts*. München: C. H. Beck. https://www.beck-shop.de/dauses-ludwigs-handbuch-eu-wirtschaftsrechts/product/396.

Deutsche Bundesregierung. 2016. : ': Deutsche Nachhaltigkeitsstrategie. Neuauflage 2016'.

'Documents – European Commission'. n. d. Accessed 17 November 2021. https://commission.europa.eu/documents_en.

'EC, 2020, SWD (2020) 176 Final Part 1/2 — European Environment Agency'. n. d. Policy Document. Accessed 21 December 2023. https://www.eea.europa.eu/policy-documents/swd--176-final-part.

Economidou, Marina, Bogdan Atanasiu, Dan Staniaszek, and Joana Maio. 2011. 'Europe's Buildings under the Microscope. A Country-by-Country Review of the Energy Performance of Buildings'. Building Performance Institute Europe. https://www.researchgate.net/publication/271200847_Europe%27s_buildings_under_the_microscope_A_country-by-country_review_of_the_energy_performance_of_buildings.

'Effort Sharing Regulation'. n. d. Text. European Commission – European Commission. Accessed 22 December 2023. https://ec.europa.eu/commission/presscorner/detail/en/qanda_21_3543.

Ekardt, Felix, Jutta Wieding, Beatrice Garske, and Jessica Stubenrauch. 2018. 'Landnutzungs- Und Düngungsbezogener Klimaschutz in Europa- Und Völkerrechtlicher Perspektive'. *ZUR*, no. 3: 143–55.

Ekkenga, Jens, Carsten Schirrmacher, and Björn Schneider. 2021. 'Ekkenga/Schirrmacher/Schneider: Offene Fragen Zur Rechtlichen Steuerung Nachhaltigen Unternehmertums'. *Neue Juristische Wochenschrift*, no. 21: 1509–13.

'Energy – European Commission'. 2023. 19 December 2023. https://energy.ec.europa.eu/index_en.

'Energy Efficiency in Buildings'. n. d. Accessed 15 October 2021. https://ec.europa.eu/info/news/focus-energy-efficiency-buildings-2020-feb-17_en,.

Epiney, Astrid. 2019. *Umweltrecht der Europäischen Union*. 4. Auflage. Baden-Baden: Nomos.

Essential EU Climate Law. 2015. Cheltenham, UK: Edward Elgar Publishing. http://www.e-elgar.com/shop/essential-eu-climate-law.

EU Technical Expert Group On Sustainable Finance. 2020. 'Taxonomy: Final Report'. https://finance.ec.europa.eu/system/files/2020-03/200309-sustainable-finance-teg-final-report-taxonomy_en.pdf.

'European Climate Law – European Commission'. n. d. Accessed 22 December 2023. https://climate.ec.europa.eu/eu-action/european-climate-law_en.

European Commission. 2019. 'Special Eurobarometer 490, Climate Change Report'. https://ec.europa.eu/clima/sites/clima/files/support/docs/report_2019_en.pdf.

European Commission. n. d. 'FAQ. What Is the EU Taxonomy Article 8 Delegated Act and How Will It Work in Practice?' https://ec.europa.eu/info/sites/default/files/business_economy_euro/banking_and_finance/documents/sustainable-finance-taxonomy-article-8-faq_en.pdf.

European Parliament Resolution of 28 November 2019 on the 2019 UN Climate Change Conference in Madrid, Spain (COP 25) (2019/2712(RSP)). 2019. https://eur-lex.europa.eu/legal-content/EN/TXT/?uri=CELEX%3A52019IP0079%2801%29.

'Energy Strategy'. n. d. Accessed 3 May 2024. https://energy.ec.europa.eu/topics/energy-strategy_en.

'EU Economy and Society to Meet Climate Ambitions'. n. d. Text. European Commission - European Commission. Accessed 3 May 2024. https://ec.europa.eu/commission/presscorner/detail/en/ip_21_3541.

'European Green Deal'. n. d. Text. European Commission - European Commission. Accessed 14 November 2021. https://ec.europa.eu/clima/eu-action/european-green-deal/european-climate-law_en.

European Parliament Resolution of 14 March 2019 on Climate Change—a European Strategic Long-Term Vision for a Prosperous, Modern, Competitive and Climate Neutral Economy in Accordance with the Paris Agreement (2019/2582(RSP)). 2019. https://eur-lex.europa.eu/legal-content/EN/TXT/?uri=CELEX%3A52019IP0217.

European Parliament Resolution of 28 November 2019 on the 2019 UN Climate Change Conference in Madrid, Spain (COP 25) (2019/2712(RSP)). 2019. https://eur-lex.europa.eu/legal-content/EN/TXT/?uri=CELEX%3A52019IP0079%2801%29.

European Parliament Resolution of 15 January 2020 on the European Green Deal (2019/2956(RSP)). 2020. https://eur-lex.europa.eu/legal-content/EN/TXT/?uri=CELEX%3A52020IP0005.

Everling, Ulrich. 1991. 'Abgrenzung Der Rechtsangleichung Zur Verwirklichung Des Binnemarktes Nach Art. 100a EWG Durch Den Gerichtshof'. *EuR*.

Falke, Josef. 2021. 'Neue Entwicklungen Im Europäischen Umweltrecht (Berichtszeitraum: 1.7.2020–12.11.2020)'. ZUR, no. 1: 57–63.

Frenz, Walter. 1997. Europäisches Umweltrecht. München: Beck.

Frenz, Walter. 2010. Handbuch Europarecht. 6: Institutionen und Politiken. Handbuch Europarecht. Erscheinungsort nicht ermittelbar.

Frenz, Walter. 2020. 'Klimaeuroparecht Zwischen Green Deal Und Corona'. *Europarecht*, no. 6: 605–21.

Frenz, Walter, and Anna-Miriam Kane. 2010. 'Die Neue Europäische Energiepolitik'. *NuR*, 464–75.

Germelmann, Claas Friedrich. 2020. 'Der Investitionsschutz Im Energie- Und Klimaschutzrecht Zwischen „European Green Deal" Und Grenzen Des Unionsprimärrechts'. *Europarecht*, no. 4: 375–407.

Grabitz, Eberhard, Meinhard Hilf, and Martin Nettesheim. 2021. *Das Recht Der Europäischen Union*. München: C. H. Beck.

Granas, Dionysios. 2013. 'Die Primärrechtlichen Grundlagen Für Die Förderung von Erneuerbaren Energien Im Europarecht'. *Europarecht*, no. 5: 619–31.

Härtel, Ines, and Johann-Christian Pielow. 2014. *Effektiv zu mehr Energieeffizienz?: Dokumentation der XV. Jahrestagung des Instituts für Berg- und Energierecht am 19. Mai 2011*. 1st ed. Richard Boorberg Verlag.

Hartmann, Moritz. 2011. 'Globaler Klimawandel Und Europäischer Rechtspluralismus'. *Europarecht*, no. 5: 636–55.

Herdegen, Matthias. 2020. *Europarecht*. 22., Überarbeitete und erweiterte Auflage. Grundrisse Des Rechts. München: C.H. Beck.

Hermelink, Andreas, and Kjell Bettgenhäuser. 2021. 'The European Commision's Renovation Wave Initiative for the Building Sector. Discussion Paper'. 53/2021. Dessau-Roßlau: Umweltbundesamt. https://www.umweltbundesamt.de/sites/default/files/medien/1410/publikationen/2021-07-21_cc_53-2021_eu-kom_renovation_wave.pdf.

Hobe, Stephan, and Michael Lysander Fremuth. 2020. *Europarecht*. 10., neu Bearbeitete Auflage. Academia Iuris Lehrbücher der Rechtswissenschaft. München: Verlag Franz Vahlen.

Horspool, Margot, Matthew Humphreys, and Michael Wells-Greco. 2018. *European Union Law*. 10th ed. Oxford New York, NY: Oxford University Press.

Hummel, Thomas, and Matthias Kolb. 2021. 'Fit for 55: So will die EU ihre Klimaziele erreichen'. Süddeutsche.de. 14 July 2021. https://www.sueddeutsche.de/politik/fit-for-55-klimaschutz-eu-1.5351072.

'Integrated National Energy and Climate Plan Pursuant to the REGULATION OF THE EUROPEAN PARLIAMENT AND OF THE COUNCIL on the Governance of the Energy Union and Climate Action, Amending Directive 94/22/EC, Directive 98/70/EC, Directive 2009/31/EC, Regulation (EC) No 663/2009, Regulation (EC) No 715/2009, Directive 2009/73/EC, Council Directive 2009/119/EC, Directive 2010/31/EU, Directive 2012/27/EU, Directive 2013/30/EU and Council Directive (EU) 2015/652 and Repealing Regulation (EU) No 525/2013'. 2022. https://energy.ec.europa.eu/system/files/2022-08/de_final_necp_main_en.pdf.

Intergovernmental Panel on Climate Change. 2019. 'Special Report on the Impacts of Global Warming of 1.5 °C above Pre-Industrial Levels and Related Global Greenhouse Gas Emission Pathways, in the Context of Strengthening the Global Response to the Threat of Climate Change, Sustainable Development, and Efforts to Eradicate Poverty'. IPCC. https://www.ipcc.ch/sr15/.

Jans, Jan H. 2003. *The European Convention and the Future of European Environmental Law*. 1., Aufl. Groningen: Europa Law Publishing.

Kahl, Wolfgang, Klaus Ferdinand Gärditz, and Reiner Schmidt. 2021. *Umweltrecht*. 12., Vollständig neu bearbeitete Edition. München: C.H.Beck.

Kanalan, Ibrahim. 2018. 'Kanalan: Rechtstreue Als Voraussetzung Für Den Anspruch Auf Die Gewährleistung Eines Menschenwürdigen Existenzminimums'. *Neue Zeitschrift Für Sozialrecht*, no. 16: 641–46.

Knodt, Michèle, and Sabine Sclacke. 2019. 'Das Governance-System Für Die Europäische Energieunion Und Für Den Klimaschutz'. *ZUR*, no. 7: 404–12.

Koch, Hans-Joachim. 2011. 'Koch: Klimaschutzrecht'. *Neue Zeitschrift Für Verwaltungsrecht*, no. 11: 641–54.

'Kurzinformation Bedeutung Der Erwägungsgründe Bei Der Umsetzung von Richtlinien (Am Beispiel Der Richtlinie 2016/943)'. 2019. PE 6–3000–020/19. Deutscher Bundestag. https://www.bundestag.de/resource/blob/628192/23bfc89c0c4ffb15489850b0558ce23f/PE-6-020-19-pdf-data.pdf.

Lamy, Christoph, and Inga Sophia Bach. 2020. 'Lamy/Bach: Die EU-Taxonomie-Verordnung Und Ihre Auswirkungen Auf Die Energiewirtschaft'. *Zeitschrift Für Das Gesamte Recht Der Energiewirtschaft*, no. 10: 348–54.

Leal-Arcas, Rafael, and Jan Wouters, eds. 2017. *Research Handbook on EU Energy Law and Policy*. Research Handbooks in European Law. Cheltenham, UK: Edward Elgar Publishing.

Löschel, Andreas, Veronika Grimm, Barbara Lenz, and Frithjof Staiß. 2021. 'Wieso Deutschland den Green Deal der EU nutzen und verstärken sollte'. *FAZ.NET*, 17 August 2021. https://www.faz.net/aktuell/wirtschaft/klima-nachhaltigkeit/wieso-deutschland-den-green-deal-der-eu-nutzen-und-verstaerken-sollte-17486210.html.

Martin Achtnicht, and Madlener Reinhard. 2014. 'Factors Influencing German Owner's Preferences on Energy Retrofits'. *Energy Policy*, no. 68: 254–63.

Meßerschmidt, Klaus. 2011. *Europäisches Umweltrecht*. 1st ed. München: C.H.Beck.

Moczko, Dominika M. 2020. 'Klimaneutrales Europa Bis 2050 – European Green Deal Und Die Fernwärme'. *Infrastruktur Recht*, no. 8: 194–96.

OECD. 2017. 'Investing in Climate. Investing in Growth',. https://www.oecd-ilibrary.org/docserver/9789264273528-en.pdf?expires=1637158414&id=id&accname=guest&checksum=9F0D139A5360DEF7137C638E21E3659F.

Oppermann, Thomas, Claus Dieter Classen, and Martin Nettesheim. 2016. *Europarecht*. 7., Völlig neu bearbeitete Edition. München: C.H.Beck.

Parikh, Jyoti et al. 2020. "Buildings". In: Climate Change 2014: Mitigation of Climate Change. Contribution of Working Group III to the Fifth Assessment Report of the Intergovernmental Panel on Climate Change. https://Www.Ipcc.Ch/Site/Assets/Uploads/2018/02/Ipcc_wg3_ar5_chapter9.Pdf.

Pechstein, Matthias, Carsten Nowak, and Ulrich Häde, eds. 2017. *Frankfurter Kommentar Zu EUV, GRC Und AEUV*. Tübingen: Mohr Siebeck.

Peeters, Marjan, and Natassa Athanasiadou. 2020. 'The Continued Effort Sharing Approach in EU Climate Law: Binding Targets, Challenging Enforcement?' *Review of European, Comparative & International Environmental Law* 29 (2): 201–11. https://doi.org/10.1111/reel.12356.

Proposal for a Directive of the European Parliament and of the Council on Energy Efficiency (Recast). 2021. https://eur-lex.europa.eu/legal-content/EN/TXT/?uri=CELEX%3A52021PC0558.

Proposal for a Regulation of the European Parliament and of the Council Establishing the Framework for Achieving Climate Neutrality and Amending Regulation (EU) 2018/1999 (European Climate Law). 2020a. https://eur-lex.europa.eu/legal-content/EN/TXT/?uri=CELEX%3A52020PC0080.

Proposal for a Regulation of the European Parliament and of the Council Establishing the Framework for Achieving Climate Neutrality and Amending Regulation (Eu) 2018/1999 (European Climate Law). 2020b. https://eur-lex.europa.eu/legal-content/EN/TXT/?uri=CELEX%3A52020PC0080.

Raux, Corinne, and Séverin Fischer. 2021. 'Testing the Application of the EU Taxonomy to Core Banking Products: High Level Recommendations'. https://www.ebf.eu/wp-content/uploads/2021/01/Testing-the-application-of-the-EU-Taxonomy-to-core-banking-products-EBF-UNEPFI-report-January-2021.pdf.

Regulation (EU) 2018/1999 of the European Parliament and of the Council of 11 December 2018 on the Governance of the Energy Union and Climate Action, Amending Regulations (EC) No 663/2009 and (EC) No 715/2009 of the European Parliament and of the Council, Directives 94/22/EC, 98/70/EC, 2009/31/EC, 2009/73/EC, 2010/31/EU, 2012/27/EU and 2013/30/EU of the European Parliament and of the Council, Council Directives 2009/119/EC and (EU) 2015/652 and Repealing Regulation (EU) No 525/2013 of the European Parliament and of the Council (Text with EEA Relevance.). 2018. *OJ L.* Vol. 328. http://data.europa.eu/eli/reg/2018/1999/oj/eng.

Saurer, Johannes. 2017. 'Saurer: Klimaschutz Global, Europäisch, National – Was Ist Rechtlich Verbindlich?' *Neue Zeitschrift Für Verwaltungsrecht*, no. 21: 1574–79.

Schlacke, Sabine, Miriam Köster, and Ass Eva-Maria Thierjung. 2021. 'Das „Europäische Klimagesetz" Und Seine Konsequenzen'. *Europäische Zeitschrift Für Wirtschaftsrecht*, no. 14: 620–26.

Schlacke, Sabine, Miriam Köster, Helen Wentzien, and Eva-Maria Thierjung. 2021b. 'Schlacke/Köster/Wentzien/Thierjung: Kursänderung Der EU: Verschärfung Der Klimaschutzziele'. *Zeitschrift Für Das Gesamte Recht Der Energiewirtschaft*, no. 1: 7–13.

Scholtka, Boris, and Laurenz Keller-Herder. 2019. 'Die Entwicklung Des Energierechts Im Jahr 2018'. *Neue Juristische Wochenschrift*, no. 13: 897–901.

Sikora, Alicja. 2021. 'European Green Deal – Legal and Financial Challenges of the Climate Change'. *ERA Forum* 21 (4): 681–97. https://doi.org/10.1007/s12027-020-00637-3.

Simon, Denys. 2001. *Le Système juridique communautaire, 3e édition.* 3e édition. Paris: Presses Universitaires de France – PUF.

Streinz, Rudolf, and Walther Michl, eds. 2018. *EUV/AEUV: Vertrag Über Die Europäische Union, Vertrag Über Die Arbeitsweise Der Europäischen Union, Charta Der Grundrechte Der Europäischen Union.* 3. Auflage. Beck'sche Kurz-Kommentare, Band 57. München: C.H. Beck.

Theobald, Christian, and Jürgen Kühling. 2021. *Energierecht.* München: C. H. Beck.

'Übermittlungsvermerk' Europäischer Rat CO EUR 13 CONCL 5 Des Für Die Betr.: Generalsekretariats Des Rates Delegationen Tagung Des Europäischen Rates (23./24. Oktober 2014) – Schlussfolgerungen'. 2014. Europäischer Rat. https://www.consilium.europa.eu/uedocs/cms_data/docs/pressdata/de/ec/145424.pdf.

Union, Publications Office of the European. 1991. CELEX1, Case C-251/91: Reference for a preliminary ruling from the Tribunal d'Instance de Beziers by order dated 27 September 1991 in the case of Roland Teulie v. Cave Cooperative Les Vignerons Puissalicon.

Urteil des Gerichtshofs (Große Kammer) vom 22. Januar 2014 Vereinigtes Königreich Großbritannien und Nordirland gegen Europäisches Parlament und Rat der Europäischen Union Verordnung (EU) Nr. 236/2012 – Leerverkäufe und bestimmte Aspekte von Credit Default Swaps – Art. 28 – Gültigkeit – Rechtsgrundlage – Eingriffsbefugnisse der Europäischen Wertpapier- und Marktaufsichtsbehörde in Ausnahmesituationen. n. d. Europäischer Gerichtshof (Große Kammer). Accessed 22 December 2023.

Heselhaus, F. Sebastian M. 2001. Abgabenhoheit der Europäischen Gemeinschaft in der Umweltpolitik: eine Untersuchung unter besonderer Berücksichtigung der Möglichkeiten und Grenzen einer Ertragshoheit der Europäischen Gemeinschaft. Hamburger Studien zum europäischen und internationalen Recht 28. Berlin: Duncker & Humblot.

Karras, Jana, and Aleksandra Novikova. 2021. 'Germany. Legal Barriers to the Decarbonisation of Dwellings Occupied by Low-Income Tenants and Financial Incentives to Overcome These Barriers'. IKEM Climate Finance Week. https://www.ikem.de/wp-content/uploads/2021/05/Brief_5_Legal-barriers-to-the-decarbonisation-of-dwellings.pdf.

Parikh, Jyoti et al. "'Buildings". In: Climate Change 2014: Mitigation of Climate Change. Contribution of Working Group III to the Fifth Assessment Report of the Intergovernmental

Panel on Climate Change'. In. https://www.ipcc.ch/site/assets/uploads/2018/02/ipcc_wg3_ar5_chapter9.pdf.

Regulation (EU) 2021/241 of the European Parliament and of the Council of 12 February 2021 Establishing the Recovery and Resilience Facility. 2021. OJ L. Vol. 057. http://data.europa.eu/eli/reg/2021/241/oj/eng.

'Smart Readiness Indicator - European Commission'. n. d. Accessed 2 May 2024. https://energy.ec.europa.eu/topics/energy-efficiency/energy-efficient-buildings/smart-readiness-indicator_en.

'Sustainable Finance'. n. d. Text. European Commission - European Commission. Accessed 2 December 2023. https://ec.europa.eu/commission/presscorner/detail/en/ip_21_3405.

Talus, Kim. 2013. EU Energy Law and Policy: A Critical Account. Oxford, New York: Oxford University Press.

Walker, Gordon, and Rosie Day. 2012. 'Fuel Poverty as Injustice: Integrating Distribution, Recognition and Procedure in the Struggle for Affordable Warmth'. *Energy Policy*, Special Section: Fuel Poverty Comes of Age: Commemorating 21 Years of Research and Policy, 49 (October): 69–75. https://doi.org/10.1016/j.enpol.2012.01.044.

Woerdman, Edwin, and Martha Roggenkamp, eds. 2015. Essential EU Climate Law. Cheltenham, UK: Edward Elgar Publishing. https://www.e-elgar.com/shop/essential-eu-climate-law.

Country Reports and their Comparative Analysis

5

5.1 Introduction

The aim of this chapter is to compare the legal and non-legal contexts of Germany and Sweden. Based on the hypothesis that Sweden is a pioneer of the energy transition, the aim is to examine whether there are legal solutions to landlord dilemmas in this country that could be transferred to the German legal system. Using the contextual method of legal comparison, it first defines the landlord-tenant dilemma in both countries and identifies barriers to energy efficiency improvement. It is anticipated that Sweden might possess solutions, potentially outside the legal sphere, due to the differing contexts of the two countries. Before analysing the replicability of Swedish experiences, it is necessary to examine the legal and non-legal contexts, as well as the legal backgrounds of both nations. The subsequent evaluation assesses the relevance of these experiences to Germany, both generally and in specific regions. Even if Swedish experiences are not directly transferable to Germany, they might still provide inspiration for innovative solutions.

It is expected that the results of this evaluation and the proposed solution will apply to other German regions—to big booming cities and the rural areas affected by the shifts related to the energy transition. Moreover, these results might apply to other EU member countries with a high share of rented dwellings, a similar climate (e.g., the Netherlands, Denmark, Austria) and face the same problem of split incentives. To compare German and Swedish legal experiences of dealing with split incentives affecting vulnerable social groups, the following questions.

J. Karras, *Legal Barriers to the Energy Modernisation of Dwellings Occupied by Low-income Tenants and Opportunities to Overcome these Barriers*, Schriftenreihe des Instituts für Klimaschutz, Energie und Mobilität,
https://doi.org/10.1007/978-3-658-44193-7_5

Following the comparative law method, the following research questions were formulated for the legal comparison and the subsequent evaluation. Due to the contextual method, the non-legal context will be considered.

1. Do the examined legal systems ensure a certain standard of living conditions, including thermal efficiency, for low-income households? If so, how are these assurances implemented?
2. Have low-income tenants encountered legal obstacles concerning the decarbonisation of residences within these legal systems?
3. Have any legal barriers been successfully addressed within these systems? If so, how?
4. Are there non-legal strategies or instruments within these legal systems that might address the landlord-tenant dilemma?

As the functional method plays a crucial role in this research, it is essential to remind that even if the German legal order faces some barriers concerning the researched topic, while Swedish law has probably overcome these barriers, it is not correct to say that Swedish law is better than German (the same relates to the researched solutions). As the possible solutions may be hidden in other areas of law as in Germany or even completely beyond the legal order, the Swedish non-legal context will also be considered.

The following chapter will answer the first three questions for each comparative country. The last question concerning other instruments for energy modernisation outside rental law will be discussed in Chap. 7 based on the results of the comparative analysis performed in the Chap. 6.

5.2 German Context

5.2.1 Non-legal Framework

Germany, as a federal republic comprising 16 states (*Länder*), possesses diverse administrative systems, legislations, and histories. According to Art. 70 (1) of the Basic Law (GG), the Länder have the right to legislate insofar as the GG does not confer legislative power on the federation. The most legal spheres related to decarbonisation are within the concurrent legislative power of the Federation (compare Art. 74 (1) Nr. 11, Nr. 24 GG).[1] Non-legal factors significantly influence the financing of decarbonisation in low-income dwellings.

For the purpose of exploring contextual differences in energy transition, this thesis compares two German regions: the metropolis of Berlin and the district of Spree-Neiße

[1] More on this matter: Knauff 2022, pp. 47–52.

in Brandenburg. Berlin, a thriving urban centre, benefits significantly from globalisation, while Spree-Neiße, a rural area, is marked by coal mining, structural change, and features like large-scale agriculture and tourism areas such as the Spreewald.[2] This thesis partially draws on results from the ESRa project, which fosters synergies between legal science and practical applications, underscoring the significance of the research topic. The regions of Berlin and Spree-Neiße were selected for their involvement in the ESRa project,[3]. providing diverse contexts for comparison. This comparison aims to ascertain if region-specific policies are essential for decarbonising the building sector or if a broader federal framework might be adequate for all regions.

The energy transition, crucial for the upcoming decades, encompasses significant social, political, economic, and technological challenges. The building sector, responsible for about 36% of CO_2 emissions,[4] is key to the success of this transition and the achievement of climate goals. However, it does not come without the risk of increasing social inequalities. The EU's commitment to a just transition acknowledges the varying impacts on regions due to their diverse backgrounds. This difference means some areas may benefit more from globalisation and the energy transition, while others, especially those reliant on outdated industries or lacking infrastructure, could face additional challenges.[5] Patterns of regional inequality, present not only in Germany but globally, may be exacerbated by major changes in the energy system. Rural areas often face higher financial burdens due to reliance on private transport, especially with increasing fuel prices. In contrast, metropolitan cities, with better infrastructure and alliances, are more likely to adapt effectively to the energy transition. The development of new business models in areas like energy transition and digitalisation tends to favour locations with existing infrastructure advantages. Areas experiencing structural changes, like Lusatia with its coal phase-out, fear a decline in living standards.[6]

Both regions are very different at first glance. They face different barriers when dealing with the decarbonisation of dwellings, especially those occupied by low-income households. The share of the rented dwellings in Berlin is very high, about 83%, while in Brandenburg, it is lower—about 48%.[7] Nevertheless, both regions deal with the same problem: how to finance the decarbonisation of dwellings, particularly those occupied by socially vulnerable groups with no access to financial assets and who cannot bear burdens related to the energetic modernisation.

---

[2] ESRa AP4, p. 10f.

[3] 'Energiewende im sozialen Raum'. n.d. 29.04.2022. https://esra-projekt.de

[4] 'In Focus: Renewable Energy in Europe'. n.d. Accessed 2 December 2023. . https://commission.europa.eu/news/focus-renewable-energy-europe-2020-03-18_en

[5] ESRa AP4.

[6] ESRa Gesamtvorhabenbeschreibung, p. 5f.

[7] 'Eigentumsquote'. n.d. Statistisches Bundesamt. Accessed 4 May 2024. https://www.destatis.de/DE/Themen/Gesellschaft-Umwelt/Wohnen/Tabellen/tabelle-eigentumsquote.html.

Berlin can be characterised by a population growth of 8.1% between 2005 and 2019, whereas a population decline is characteristic of the Spree-Neiße region. Between 2005 and 2019, the population in Spree-Neiße fell by 16.9%, while employment density and the unemployment rate have steadily risen.[8] Meanwhile, Berlin achieved the highest GDP growth of all federal states at 3.0% in 2019. The two regions have markedly different economic structures. In the last decade, the service sector has expanded enormously in Berlin. The start-up scene has significantly developed in the capital of Germany in recent years. In no other federal state were so many businesses registered in 2019. Additionally, the city is distinguished by its universities and research institutes, and tourism. Most of Berlin's businesses (75,653) are small companies with a turnover of less than 50,000 euros per year.[9]

The population of Berlin is notably heterogeneous, with city districts varying significantly in socio-economic, social, cultural, and political factors, thus forming distinct spatial distribution patterns.[10] The city's demographic profile is relatively young and highly international. In 2019, nearly one in five residents in Berlin was a foreign national (19.24%).[11] In terms of immigration, there was a notable influx of 103.9 individuals per 10,000 inhabitants from abroad in 2018.[12] However, internal migration trends present a contrasting picture: 23.15% of the inhabitants relocated from Berlin to other German districts in the same year. A driving factor behind this trend is the escalating cost of housing in Berlin, coupled with the increasing demand for available building plots. In 2018, the prices for building plots for owner-occupied homes in Berlin ranged between 300 and 500 euros per m^2,[13] whereas in Spree-Neiße, even in 2020, prices remained below 50 euros per m^2.[14]

Berlin's industrial development is based on longstanding expertise in the metal industry, electrical engineering, as well as mechanical and vehicle engineering, including renewable energies. Together with the chemical-pharmaceutical industry, these sectors are among the most vital in Berlin in terms of turnover and employment. Nevertheless, almost 40% of Berlin's employees worked in large companies in 2019.[15] Nevertheless, both regions are confronted with an identical challenge: financing the decarbonisation of

[8] ESRa AP3, p. 7.

[9] 'Eigentumsquote'. n.d. Statistisches Bundesamt. Accessed 4 May 2024. https://www.destatis.de/DE/Themen/Gesellschaft-Umwelt/Wohnen/Tabellen/tabelle-eigentumsquote.html; Karras/ Novikova 2021, p. 1; ESRa AP3, p. 7.

[10] ESRa P4.2., p. 26.

[11] BMI 2019; ESRa P4.2., p. 28.

[12] Ibid.

[13] ESRa P4.2., p. 31.

[14] Gutachterausschuss für Grundstückswerte in den Landkreisen Spree-Neiße und Oberspreewald-Lausitz 2021, p. 22.

[15] Ibid.

dwellings, particularly those inhabited by socially vulnerable groups. These groups lack access to financial assets and are unable to shoulder the costs associated with energy-efficient modernisation.

In the Spree-Neiße district, a significant portion of the gross domestic product is attributed to the manufacturing industry, which encompasses sectors like mining, energy and water supply, and the broader energy industry. The manufacturing sector, especially mining, plays a pivotal role in Spree-Neiße's economy. Despite the initiation of some energy transition projects, the district's economic structure is still heavily reliant on fossil fuels, particularly the brown coal industry.[16] The impending energy transition and the closure of coal power plants are expected to lead to job losses in the region. Conversely, in Berlin, companies generally perceive the energy transition as an opportunity for growth and new partnerships. The phase-out of brown coal and structural changes pose significant challenges for small and medium enterprises in Spree-Neiße. Nevertheless, some regional companies are gradually shifting away from fossil fuel dependency. For instance, Vestas Blades has been manufacturing rotor blades for wind turbines for approximately 19 years. This transition is also evident within the fossil fuel-based companies themselves. The economic structure of Lusatia, and particularly Spree-Neiße, is undergoing a transformation, moving away from mining and brown coal towards climate-neutral business models and processes, aiming to maintain its status as an energy region in the future.[17]

The Leibniz Institute for Research on Society and Space, in collaboration with Germanwatch e. V., has undertaken an analysis of the causes and explanatory approaches for region-specific attitudes towards energy transition within the model regions of the ESRa project. This involved conducting a series of interviews with stakeholders and energy transition experts to capture the patterns of these attitudes. Both regions acknowledged the relevance and necessity of climate protection and energy transition. However, the prevailing attitude in Berlin appears more optimistic than in the Spree-Neiße district. This variance can be attributed to differences in the perceived positive or negative external impacts of the transformation sectors. Berlin's focus is predominantly on emission reduction, achieving climate targets, and human health. In contrast, Spree-Neiße's emphasis lies on the effects on jobs, economic development, and regional identity, all of which significantly influence the residents' perception of the energy transition.[18]

Furthermore, the researchers analysed the local daily newspapers *Lausitzer Rundschau* and *Tagesspiegel Berlin*. Their findings indicate that these media outlets present differing perspectives on the transformation sectors. *Lausitzer Rundschau*'s coverage primarily highlights structural change and the expansion of renewable energies, while *Tagesspiegel Berlin* tends to discuss the energy transition in the context of transport

---

[16] ESRa AP3, p. 8.

[17] Ibid.

[18] ESRa P1.2, p. 6.

transition. Intriguingly, only *Lausitzer Rundschau* addressed the impact of the energy transition on housing development,[19] a topic of even greater significance for Berlin given its high proportion of rented dwellings.

It is essential to understand these fundamental perceptual differences in the model regions and to take them into account when devising solutions for the landlord-tenant dilemma. A positive outlook towards energy transition could lead to increased willingness to invest in the decarbonisation of dwellings in the future. Conversely, perceptions associated with fears of negative consequences, such as job loss and loss of regional identity, may hinder the readiness to invest in the energy modernisation of dwellings, unless the benefits are directly experienced by the concerned parties.

5.2.2 German Legal Framework

5.2.2.1 General Overview

As the average reader is likely familiar with the German legal context, only a brief overview is provided here, aiming to furnish those with limited knowledge of the German legal system with some important details. Continental European law, of which German law is a part, is predominantly statutory in nature: law is manifested through statutes, and these statutes embody the essence of the law. This continental European, and specifically German, understanding of law is fundamentally positivistic. Statutes are not perceived as exceptions or deliberate deviations from the law; rather, they are the law itself.[20] In this framework, there is no law independent of the statute.[21]

However, there are notable exceptions to this basic principle, particularly in Germany's treatment of the statutes from the two totalitarian regimes of the 20th century: the National Socialist dictatorship in the Third Reich and the Socialist dictatorship in the GDR. This approach was articulated by the German legal philosopher *Gustav Radbruch* and later adopted by the courts.[22] He stated: "Der Konflikt zwischen der Gerechtigkeit und der Rechtssicherheit dürfe dahin zu lösen sein, daß das positive, durch Satzung und Macht gesicherte Recht auch dann den Vorrang hat, wenn es inhaltlich ungerecht und unzweckmäßig ist, es sei denn, daß der Widerspruch des positiven Gesetzes zur Gerechtigkeit ein so unerträgliches Maß erreicht, daß das Gesetz als „unrichtiges Recht" der Gerechtigkeit zu weichen hat."[23]

---

[19] Ibid, p. 67.

[20] In contrast to classical common law, Kischel 2015, p. 392.

[21] Ibid.

[22] Ibid, p. 393.

[23] Radbruch 1950, p. 353; The conflict between justice and legal certainty should be resolved in such a way that the positive law, secured by statute and power, takes precedence even if it is substantively unjust and inexpedient, unless the contradiction of the positive law with justice reaches

This reflects the influence of natural law, suggesting that certain legal norms, even if officially enacted, might not be regarded as law if they contravene fundamental legal principles that exist independently of any state recognition.[24] Despite this influence of natural law on the German legal system, its relevance to the present work is minimal and, as such, will not be explored in detail.

Understanding the centrality of statutes in the German legal system is crucial for comparative analysis, particularly for exploring the interplay between legal and non-legal contexts. [25] It can be presumed that the answers to the research questions will predominantly, though not exclusively, be found in statutes, as they are intrinsically regarded as law in Germany. Conversely, this understanding also cautions against focusing solely on statutes or written law when researching a foreign country, as it may not necessarily adopt the same positivist approach.

In addition to statute (*Gesetz*) and customary law, which plays rather a relatively minor role in the German legal system, there are two other notable written sources of law: bylaws (*Satzungen*) and collective agreements.[26] The latter will not be further examined in this thesis due to its limited scientific relevance.[27] However, bylaws warrant closer examination. The key distinction between bylaws and ordinances (*Verordnungen*) lies in their issuance by self-governing bodies to regulate their internal affairs, separate from the executive and legislative branches. An example of bylaws includes a municipality's development plans (*Bebauungspläne*). Although these self-governing bodies are subordinate to the state, the distinct categorisation of bylaws as a source of law merely reflects a formal characteristic of the German state organisation.[28]

The role of fundamental rights in the German legal system is difficult to overestimate. The fundamental rights here reflect the value system thinking rooted in Weimar constitutional law.[29] German jurists interpret legal facts through the lens of this value system. Before delving into the Swedish approach in detail, it is imperative to clarify this aspect of the German legal framework. Additionally, it is essential to reiterate that in comparative law analysis, other legal systems should be considered independently of one's own legal value system thinking.

such an intolerable degree that the law, as "incorrect law" must give way to justice (own translation), .

[24] Kischel 2015, p. 393.

[25] Ibid, p. 392.

[26] Kischel 2015, p. 401.

[27] The controversial legal source of "general principles of law", which are only very rarely mentioned at all in the German legal context, and are then regarded mostly as a part of the problematic judge law (Kischel 2015, p. 401), are also deliberately disregarded here.

[28] Kischel 2015, p. 401.

[29] Herdegen in: Maunz/Dürig 2021, GG Art. 1 Rn. 60.

A catalogue of fundamental rights is integral not only to the value system but also to the cultural system, which is inextricably linked to the democratic, civil, and constitutional state governed by the rule of law.[30] *Carl Smitt* describes the value-creating significance of fundamental rights as a fundamental political decision of the German nation.[31] The basis of the German value system is the concept of human dignity establishedin Art. 1 (1) GG, which is further developed by the supranational human rights under Art. 1 (2) GG. These human rights are given comprehensive effect by Art. 1 (3) GG affirming their direct applicability. Complete protection of fundamental rights in Germany is guaranteed by Art. 2 (1) GG, Art. 3 (1) GG and Art. 19 (4) GG, which secure right of freedom, equality, and procedural protections thereof.[32]

Another important aspect is the distribution of competences across different governmental levels. As a federal state, as defined in Art. 20 (1) GG, Germany's distribution of competences between the Federation (*Bund*) and the *Länder* is fundamentally influenced by its federal structure.[33] In this context, Art. 30 GG stipulates that the exercise of state power and the fulfilment of state tasks is a matter of the Länder, except where the GG provides or permits otherwise.[34] The German Constitution does not allocate state tasks and powers based on specific subject areas, resulting in neither the Federation nor the *Länder* having exclusive legislative, executive, or judicial authority within specific domains. Instead, state powers are categorised into four domains: foreign affairs, legislative, administrative, and judicial. The competences within these domains are then allocated between the Federation and the *Länder* according to specific subject areas.[35]

The legislative acts analysed in subsequent sub-chapters are part of federal legislation and are enacted based on the legislative competences of the Federation. Fundamental rights following from the constitutions of the *Länder* are not addressed in this thesis, as they do not extend a broader scope of protection than the GG in relation to the research questions. The allocation of relevant administrative competences between the Federation and the *Länder* will be elucidated later at the appropriate juncture.

5.2.2.2 Is any Level of Living Conditions, Including Thermal Efficiency, Guaranteed by the Legislator in Germany?

The assessment of the question related to the guaranteed level of living provides an overview of the minimum living conditions that are ensured by the legislator. This is crucial for defining the level of security for energy-poor households and examining whether they have an enforceable claim to cover the expenses for energy modernisation of their

---

30 Ibid.

31 Schmitt 1928, p. 162f.

32 Herdegen in: Maunz/Dürig 2021, GG Art. 1 Rn. 60f.

33 Katz/Sander 2019, Rn. 259.

34 Grzeczick in: Maunz/Dürig 2021, Art. 20 Rn. 67f.; Kment in: Jarass/Pieroth 2020, Art. 30 Rn. 1.

35 Grzeczick in: Maunz/Dürig 2021, Art. 20 Rn. 67f.

dwellings. It is important to note that there is no legal definition of energy poverty in German legislation. The definition formulated in Sect. 3.1.1. is intended to guide this discussion. According to this definition, *Energy poverty is characterised by the inability to either keep the house adequately warm or to pay utility bills in the last 12 months, or both; or reliance on social (publicly funded) assistance that covers energy bills to a reasonable extent, coupled with the necessity to conserve energy to avoid exceeding the amount covered by public funds; or challenges in maintaining adequate warmth in the household without sacrificing other needs, alongside difficulties in making ends meet, even when the individual does not receive (publicly funded) social aid.*

In Germany, the matter of a certain level of living conditions guaranteed by the legislator is intimately connected to the fundamental right of human dignity.[36] According to Art. 1 (1) GG, human dignity is inviolable, and all state authorities are duty-bound to protect it. The German Constitution not only entitles individuals to defend against state interferences but also obliges the state to actively protect human dignity.[37] The guarantee of a dignified minimum subsistence level for everyone follows from the right to human dignity combined with the welfare state principle (Art. 20 (1) GG).[38] The state is obliged to support citizens who cannot sustain themselves due to certain circumstances, such as difficult life situations or physical or mental impairments. However, as per the consistent jurisprudence of the Federal Constitutional Court, the entitlement to protection covers only the basic conditions necessary for a dignified existence.[39] The welfare state principle under Art. 20 (1) GG mandates the legislature to ensure a minimum subsistence level that is worthy of human dignity. Additionally, the legislature has the discretion to determine the amount of this minimum subsistence level.[40] According to the Federal Constitutional Court, this requirement necessitates concrete and regular updates by the legislature. The amount should be adjusted to reflect the current level of development of the commonwealth and the prevailing living conditions, within the legislative parameters.[41]

The determination of the minimum subsistence level is a topic of active discussion in the literature,[42] which aligns with case law in emphasizing that the subsistence minimum

[36] Kunig/Kotzur in: v. Münch et al. 2021, Art. 1 GG Rn. 44.

[37] Graßhof in: Dolzer et al. 2004, GG Art. 1 Abs. 1, Nr. 10,3; BVerfG, Beschluss des Ersten Senats von 11. März 2003 – 1 BvR 426/02., Rn. 26; Graßhof in: Dolzer et al. 2004, Art. 1 Abs. 1, Nr. 10,6; BVerfGE 40, 121 (133), BVerfGE 125, 175 (222).

[38] Grzeczick in: Maunz/Dürig 2021, GG Art. 20 Rn. 23; BVerfG, Urteil des Ersten Senats vom 07. Juni 2005 –1 BvR 1508/96-,Rn. 48–49.

[39] BVerfG, Urteil des Ersten Senats vom 09. Februar 2010 – 1BvL 1/09-, Rn. 110; BVerfGE 132, 134 (159); BVerfGE 137, 43 (72).

[40] BVerfGE 125, 175 (222); BVerfGE 35, 202 (246).

[41] BVerfGE 125, 175 (222).

[42] Kunig/Kotzur in: v. Münch et al. 2021, Art. 1 GG Rn. 44f.; Grzeczick in: Maunz/Dürig 2021, GG Art. 20 Rn. 25; Herdegen in: Maunz/Dürig 2021, Art. 1 GG Rn. 121.

is contingent upon the specific situation and context.[43] However, Art. 1 (1) GG does not oblige the state to reduce disparities in in wealth; rather, it impacts the lower threshold as a minimum standard, not an optimum.[44] This minimum is subject to changing circumstances. Generally, this right does not entitle one to demand specific measures to establish a particular status of social security or a certain degree of social equality beyond the minimum subsistence level.[45] Although issues surrounding human dignity in the context of a minimum subsistence level are highly sensitive and robustly protected by the state, the Federal Constitutional Court (BVerfG) consistently underscores the legislative leeway in its rulings.[46] Nevertheless, the BVerfG regularly requires eliminating deficits in the social security systems and, in some cases, establishes extensive guidelines for the legislature. In one of its rulings, the BVerfG required the legislature to evaluate all necessary expenses for existence in a transparent and suitable procedure, with the need for regular review and development.[47] In another ruling, the BVerfG considered the basic needs under the Hartz IV social benefits system still compatible with the fundamental right to guarantee a subsistence minimum in human dignity under Art. 1 (1) GG, in conjunction with the principle of the welfare state under Art. 20 (1) GG.[48]

There is a consensus in both literature and case law that the welfare state principle, as outlined in Art. 20 (1) of the GG, does not, in itself, create subjective rights to social benefits.[49] However, individuals do have a claim to those social benefits deemed necessary to lead a life worthy of human dignity, as articulated in Art. 1 (1) in conjunction with Art. 20 (1) GG.[50] This claim is designed to ensure that every individual receives assistance with the material conditions indispensable for their physical existence and a minimum level of participation in social, cultural, and political life.[51]

The German legislator distinguishes essential security benefits according to different needs. The foundational entitlement includes benefits to secure subsistence for everyday needs as per §20 SGB II. The individual in need is also entitled to benefits for additional needs according to §21 (2–5), §23 No. 2–4 SGB II (primarily related to health and disability issues), benefits for the actual reasonable expenses for accommodation and heating

[43] Kunig/Kotzur in: v. Münch et al. 2021, Art. 1 GG Rn. 44; Herdegen in Dürig et al. 2021, Art. 1 Rn. 121; Graßhof in: Dolzer et al. 2004, Art. 20 Abs. 1 Nr. 194; Kanalan 2018, p. 642.

[44] Kunig/Kotzur in: v. Münch et al. g 2021, Art. 1 GG Rn. 45;

[45] Grzeczick in: Maunz/Dürig 2021, GG Art. 20 Rn. 25.

[46] Kunig/Kotzur in: v. Münch et al. 2021, Art. 1 GG Rn. 45; Grzeczick in: Maunz/Dürig 2021, GG Art. 20 Rn. 25.

[47] BVerfGE 125, 175 (225); Kunig/Kotzur in: v. Münch et al. 2021, Art. 1 GG Rn. 45.

[48] BVerfGE 137, 34 (73); Kunig/Kotzur in: v. Münch et al. 2021, Art. 1 GG Rn. 45.

[49] So BVerfGE 27, 253 (283); BVerfGE 82, 60 (80), BSGE 155, 115 (120); Wittreck in: Dreier 2015, GG Art. 20 Rn. 45.

[50] BVerfGE 40, 121 (133); 45, 187 (228); Wittreck in: Dreier 2015, GG Art. 20 Rn. 45.

[51] Wittreck in: Dreier 2015, GG Art. 20 Rn. 45; BVerfGE 125, 175 (223); BVerfGE 80, 367 (374).

as specified in §19 (1) s. 3 SGB II. A lump sum for decentralised hot water supply is provided under §21 (7) SGB II. Moreover, individuals in need are entitled to benefits for ongoing, not merely one-time, needs that are unavoidable in the individual case, as per §21 (6) SGB II.[52]

Furthermore, a loan may be granted for a one-time need that is unavoidable under the circumstances and typically covered by everyday needs but is not, or not sufficiently, covered in the individual case, as outlined in §24 (1) SGB II. There is also an entitlement to specific one-time allowances, for example, for the initial equipment of the dwelling with household appliance under §24 (3) No. 1 SGB II. The state also provides support for moving-in and moving-out costs and rent deposits, as per §22 (6) SGB II. Additionally, the costs for education and social inclusion of children, adolescents, and young adults are recognised and provided for by the German legislature, as stated in §28 II SGB.[53]

The BVerfG conducts an evidence test to determine whether the benefits are wholy inadequate.[54] This involves examining whether the legislature has met the objective of ensuring a dignified existence in a manner that does not violate Art. 1 (1) GG in conjunction with Art. 20 (1) GG. The examination includes checking if the legislature has selected a calculation method within its discretion that is generally suitable for assessing the subsistence minimum; whether it has determined the necessary facts in a fundamentally complete and accurate manner; and finally, whether it has remained within the bounds of what is justifiable in all calculation steps, ensuring that this calculation does not violate the chosen method and its structural principles.[55]

The statutory entitlement to social benefits must be structured to consistently cover the basic needs of each affected individual. This entitlement is contingent on society's perception of what is necessary for a dignified existence, the concrete life situation of the person requiring assistance, and the respective economic and technical circumstances. The legislature is tasked with specifically determining these entitlements. The principle of the welfare state in Art. 20 (1) GG obligates the legislature to understand social reality in a manner that is timely and realistic with regard to guaranteeing a minimum subsistence level worthy of a human being. For instance, in a technological society, the necessities differ from those in the past.[56] This implies that a person in need of assistance should be entitled to a real opportunity to acquire a personal computer or smartphone, which were once considered luxuries. The legislature bears the responsibility for the necessary

[52] BVerfG, Beschluss des Ersten Senats vom 23.08.2014 – 1 BvL 10/12-, Rn. 5.

[53] Ibid.

[54] The BVerfG assumed the benefits as objectively inadequate in the case of the Asylum Seekers' Benefits Act (Asylbewerbungsleistungsgesetz), BVerfG, Urteil des Ersten Senats vom 18. Juli 2012 – 1 BvL 10/10 -; BVerfGE 125, 175 (226); Spitzlei 2017, p. 165.

[55] BVerfGE 125, 175 (226); Wittreck in: Dreier 2015, GG Art. 20 Rn. 45.

[56] SG Aachen, Urt. v. 20.7.2011 – S 5 AS 475/11, BeckRS 2012, 66953.

evaluations and concretisation of the entitlement to social benefits in terms of facts and legal consequences in this context.[57]

The German legislature has not made any specific decisions regarding the entitlement to energy-efficient housing for individuals requiring assistance. Living in energy-efficient housing is not yet perceived as a given in today's society. It is anticipated that there will be a shift in the social understanding of energy-efficient housing as part of the minimum subsistence level, which will eventually be encompassed by social assistance to ensure a dignified existence for individuals in need.

Through the so-called Social Protection Packages I to III, the German legislator temporarily simplified access to essential benefits for job seekers under SGB XII until 31 December 2021, as a response to the Covid-19 crisis.[58] In particular, the usual assessments of the reasonableness of housing rents were temporarily suspended.[59] Typically, what constitutes reasonable housing costs is determined by local guidelines set by cities and municipalities. However, during the Covid-19 crisis, substantial costs were covered to prevent individuals who lost their jobs from relinquishing their flats due to costs deemed unreasonable by existing guidelines.[60] This meant that individuals residing in energy-efficient, and thus often more expensive, flats who were affected by the Covid-19 crisis and lost their jobs, were temporarily entitled to social benefits covering their rental costs. Nonetheless, this temporary measure did not establish a precedent to shift case law or compel the legislator to re-evaluate what is considered an essential security measure necessary to guarantee human dignity.

Additionally, the German legislator has established a housing allowance for a specific group of people, aimed at economically securing adequate and family-friendly housing (§1 (1) WoGG—Housing Allowance Act). The housing allowance is granted either as a subsidy towards the rent (rent subsidy) or towards the encumbrance (encumbrance subsidy—Lastenzuschuss) for a self-used rented apartment (§1 (2) WoGG). The groups of individuals eligible for a housing allowance are outlined in §3 WoGG. It is important to note, however, that this housing allowance is not designed to support applicants in difficult financial circumstances and does not take such circumstances into account. Rather, the Housing Allowance Act focuses exclusively on income and definitively regulates the deductions from income when determining eligibility for this assistance. Its objective is purely economic,[61] supporting low-income individuals who are not entitled to other forms of social assistance, as per §7 WoGG, such as benefits for job seekers. This allowance does not guarantee any level of energy efficiency in the dwellings of its recipient.

---

[57] Ibid.

[58] BT Drucksache 19/26542.

[59] BT Drucksache 19/18107.

[60] BT Drucksache 19/18107.

[61] VG Greifswald Urt. V. 22.9.2020 – 2 A 428/20, BeckRS 2020, 27040.

In summary, the German legislator guarantees only the subsistence minimum in human dignity under Art. 1 (1) GG, in conjunction with the welfare state principle under Art. 20 (1) of the GG. The scope of this guaranteed subsistence minimum is mainly detailed in social legislation. However, this guarantee does not extend to a specific level of thermal efficiency in housing. Consequently, energy-poor households cannot derive an enforceable claim to cover the costs of energy modernisation of their dwellings.

A further question arises in this context regarding the scope of the guaranteed minimum living conditions in the German legal order, specifically the understanding of adequate housing. According to contemporary interpretations, adequate housing entails more than just having a roof over one's head. It also implies a certain standard of housing, including facilities such as energy and sanitation. The GG does not provide a specific right to housing.[62] Art. 13 (1) GG merely states that the home is inviolable and obliges the state to refrain from intervening within its protective scope. It entitles individuals to defend against state intervening into the home.[63] Art. 13 GG, however, does not establish a legal claim for adequate housing against the state, nor does it oblige the legislator to design a "social" tenancy law.[64] The guiding principles for the legislature in this regard stem from the welfare state principle (Art. 20 (1) GG) and the social obligation of property according to Art. 14 (2) GG.[65] Yet, these provisions merely determine the state's objectives.[66]

However, some Länder constitutions do at least mention a right to adequate housing, such as Art. 28 (1) of the Berlin Constitution (BlnVerf),[67] Art. 47 (1) of the Brandenburg Constitution (BnbVerf), and Art. 15 of the Thuringian Constitution. the wording of these legal norms, like the Berlin Constitution stating: "*Jeder Mensch hat das Recht auf angemessenen Wohnraum*" ("Everyone has the right to adequate housing"), Art. 28 (1) S. 1 BlnVerf, appears to confer a subjective right. Yet, the interpretation of these provisions reveals that they do not provide an individual claim to housing but rather define an objective of the Land.[68] Some Constitutions i.e., the Constitution of Brandenburg, explicitly regulate housing supply and promotion as state objectives,[69] including measures such as social housing construction, tenant protection, and rent subsidies within its powers (see Art. 47 (1) BnbVerf). According to the BlnVerf, the Land is tasked with promoting the

---

[62] BT WD3 – 3000 – 120/19, p. 3; Kunig/Berger in: Münch et al. 2021, Art. 13 Rn. 14.

[63] Kunig/Berger in: Münch v 2021, Art. 13 Rn. 14.

[64] BVerfGE 18, 121 (132); Kunig/Berger in: Münch v 2021, Art. 13 Rn. 14; Hermes in: Dreier 2015, Art. 13 Rn. 124.

[65] Kunig/Berger in: Münch et al. 2021, Art. 13 Rn. 14.

[66] BT WD3 – 3000 – 120/19, p. 3.

[67] Ibid.

[68] BT WD3 – 3000 – 120/19, p. 3; Derleder 2009, p. 616.

[69] BT WD3 – 3000 – 120/19, p. 3; further examples: Art. 17 (1) Verf. M-V; Art. 29(2) VerfNRW; Art. 40(1) S-A LVerf.

creation and maintenance of adequate housing, particularly for low-income individuals (Art. 28 (1) S. 1 BlnVerf).

The welfare state principle in Art. 20 (1) GG is reflected in Germany's social legislation. §1 (1) of the Social Code I (SGB I) states that social legislation aims to shape social services, including social assistance, to realise social justice and security. Its objectives include ensuring a dignified existence, creating equal opportunities for personal development (particularly for young people), protecting and promoting the family, enabling the earning of a basic income through chosen activities, and providing protection or compensation for substantial life burdens, also by encouraging self-help. §2 (1) GG specifies that social rights derived from the Social Codes [70] can be claimed only as detailed in the provisions of the respective sections of the SGB I. The legal reservation (*Gesetzesvorbehalt*) applies to legal relationships under the SGBs, meaning rights and obligations in social assistance are created, established, amended, or revoked only as prescribed or permitted by law, as per §31 SGB I. If benefits providers have discretionary powers in deciding on social benefits, they must exercise this discretion in accordance with the authorisation and adhere to legal limits, as stated in §39 (1) SGB I.

The first question regarding these legal provisions is whether the SGB provides any entitlement to a certain level of adequate housing for low-income households. As previously mentioned, social legislation contains numerous norms defining standard and special needs for securing subsistence, such as §§ 20 et seq. SGB II. SGB II recognises heating costs as part of everyday needs (§22 (1) SGB II) but only to the extent of actual reasonable expenses. If a household moves to another dwelling and the expenses for accommodation and heating increase beyond what the benefit provider deems necessary, only the previous expense amount is recognised, according to §22 (1) s. 2 SGB II. Expenses for accommodation and heating that exceed the reasonable amount for the individual case may be recognised as needs for up to six months, provided it is not possible or reasonable for the person or household to reduce expenses by moving to a cheaper dwelling or by other means. This recognition is not demanded if such a reduction would be uneconomical compared to the benefits provided for moving to another dwelling, as per §§22 (1) s.3, s. 4 SGB II. These provisions indirectly discourage low-income households from moving to better housing, as rents are typically higher, even if heating costs are lower due to improved energy efficiency, due to the risk that their accommodation expenses will not be recognised as reasonable.

The SGBs do not explicitly define "reasonable costs". This term represents an indeterminate legal concept (*unbestimmter Rechtsbegriff*), serving as a guideline for the executive's strict legal adherence and is generally accepted under German constitutional law. The nuances associated with indeterminate legal terms are not the focus here. However, it is crucial to note that the executive is responsible for interpreting these indeterminate

[70] There are twelve Social Codes in Germany.

legal concepts.[71] Firstly, it is the executive's duty to interpret and apply these concepts, providing evaluative concretisation of the programme set by legal norms. For instance, the executive cannot defer necessary measures and decisions due to unresolved legal questions or lack of judicial guidance on an indeterminate legal term. Secondly, the interpretation and application of these concepts are fully subject to judicial review,[72] meaning the executive must independently determine the adequacy of housing costs. For judicial review, the social courts have developed a process to assess the appropriateness of housing costs, which involves several steps.[73]

First, the dwelling size rented by a low-income household in need must be determined.[74] At this stage, the number of square meters of the dwelling specifically concerned in the dispute case must be determined. This size is based on the dimensions recognised for individuals entitled to social rented housing, as previously defined under §5 (2) of the Housing Binding Act (*Wohnungsbindungsgesetz*)[75] and its implementing provisions under Land law. Following the repeal of the *Wohnungsbindungsgesetz*, the reasonable dwelling size is now derived from §10 of the Act on Promoting Social Housing (WoFG). Under this norm, the *Länder* are empowered to set specific size limits for subsidised rental housing based on appropriateness principles. Consequently, the *Länder* issue guidelines detailing the suitable dwelling sizes.[76] In Berlin, this is governed by the Implementing Regulations for the Granting of Benefits under §22 of SGB II and §§35 and 36 of SGB

71 Geis in: Schoch/Schneider 2021, §40 VwVfG, Rn. 137.

72 Schönenbroicher in: Mann/Sennekamp/Uechtritz 2019, VwVfG §40 Rn. 83 et seq.; Wolff in: Sodan/Ziekow 2018, §114 VwGO Rn. 303; BVerwGE 15, 207 (208).

73 *Hessisches LSG,* Beschluss vom 28. 3. 2006 – L 7 AS 122/05 ER Lind L 7 AS 121/05 ER -, NZM 2006, 595; Hessisches *LSG,* Beschluss vom 24. 4. 2006 – L 9 AS 39/06 ER; *LSG Baden-Württemberg,* Beschluss vom 25. 1. 2006 – L 8 AS 4296/05 ER-B; *LSG Baden-Württemberg,* Beschluss vom 23. 3. 2006 – L 13 AS 510/06 ER-B; *LSG Nordrhein-Westfalen,* Beschluss vom 24. 8. 2005 – L 19 B 28/05 AS ER; BSG, Urteil vom 7.11.2006 – B 7b AS 16/06 R, NZS 2007, 547.

74 BSG, Urteil vom 7.11.2006 – B 7b AS 16/06 R, NZS 2007, 547.

75 The revision of Art. 72f. GG by the Federalism Reform Act of 28 August 2006 (BGBl I p. 2034) transferred some areas of housing legislation to the competence of the Länder. As per Art. 125a (1) of the GG, the Law on Securing the Purpose of Social Housing (Gesetz zur Sicherung der Zweckbestimmung von Sozialwohnungen—WoBindG) has continued to be applicable as federal law since 1 September 2006. However, it can now be superseded by Land law. Since 31 August 2006, the federal legislature no longer possesses the legislative authority to amend the WoBindG. To date, not all Länder have implemented their own regulations in the area of housing, 'Gesetz zur Sicherung der Zweckbestimmung von Sozialwohnungen—WoBindG'. n.d. 04 September https://beck-online.beck.de/Dokument?vpath=bibdata%2Fges%2Fwobindg%2Fcont%2Fwobindg.htm%26anchor=Y-100-G-WOBINDG2021.

76 BSG, Urteil vom 7.11.2006 – B 7b AS 16/06 R, NZS 2007, 547.

XII (AV-Wohnen),[77] and in Brandenburg, by the Administrative Regulation on the Housing Promotion and Housing Bond Act (VV-WoFGWoBindG).[78] Therefore, the social court must verify whether the size of the dwelling rented by the affected household falls within the size limits recognised under Land law as per the WoFG.[79]

Next, the standard of the dwelling is considered. Housing costs are deemed appropriate only if the dwelling meets basic standards in terms of furnishings, location, and building structure, and does not reflect a high-end living standard. The courts employ the so-called "product theory", suggesting that the subsidised dwelling must fall within the lower housing segment in the spatial district to be considered appropriate in size. However, the cost burden to the social benefits provider is the primary concern, making individual factors like furnishings and location irrelevant if they do not result in unreasonably high costs to the state.[80]

Neither social benefit providers nor social courts take into account the energy efficiency of a dwelling when determining housing adequacy. An important question arises as to whether the German legislator has established legally binding energy efficiency standards for rented dwellings, which can be claimed by tenants regardless of their income level or whether their rent is subsidised. To address this, a brief exploration of the overall characteristics of the housing market in terms of energy efficiency is necessary for understanding the challenges in decarbonising dwellings.

The modern German housing system is predominantly market-based, relying heavily on property ownership and private investments. However, differences still exist between the housing markets of former West and East Germany, stemming from divergent developments under two distinct socio-political systems.[81] As of the end of 2019, Germany had approximately 42.5 million residential units in both residential and non-residential buildings.[82] By the end of 2018, natural gas was the primary heating source in about 52% of dwellings, while approximately 23.5% of private households used oil for heating. Oil was also utilised for hot water supply in 20% of homes.[83] Consequently, residential

---

[77] Ausführungsvorschriften zur Gewährung von Leistungen gemäß §22 des Zweiten Buchs Sozialgesetzbuch und §§35 und 36 des Zwölften Buches Sozialgesetzbuch (AV-Wohnen) vom 02. Februar 2021 (Abl. S.3727f.).

[78] Verwaltunsgvorschrift zum Wohnraumförderungs- und Wohnungsbindungdgesetz (VV-WoFGWoBindG) vom 27. November 2017 (Abl./17, [Nr.49], S. 1122.

[79] BSG, Urteil vom 7.11.2006 – B 7b AS 16/06 R, NZS 2007, 547.

[80] Ibid.

[81] Noll/ Weick 2014, p. 1.

[82] 'Wohnungsbestand Ende 2019: 42,5 Millionen Wohnungen'. n.d. Statistisches Bundesamt. Accessed 23 December 2023. https://www.destatis.de/DE/Presse/Pressemitteilungen/2020/07/PD20_281_31231.html.

[83] 'Knapp ein Viertel der Haushalte heizt die Wohnung 2018 immer noch mit Öl'. n.d. Statistisches Bundesamt. Accessed 23 December 2023. https://www.destatis.de/DE/Presse/Pressemitteilungen/2019/10/PD19_N002_129.html.

buildings are responsible for approximately 70% of all CO_2 emissions in the building sector, with 122 million tonnes of CO_2 equivalents emitted in 2019. Although the building sector has managed to reduce its greenhouse gas emissions by almost 42%, achieving the target of only 70 million tonnes by 2030 remains a challenge.[84] Approximately 70% of the final energy consumption in buildings is attributed to room heating and hot water supply, with the decline in CO_2 emissions being primarily temperature-related.[85]

In November 2020, the Building Energy Act (Gebäudeenergiegesetz—GEG) came into effect, consolidating the Energy Savings Act, the Energy Savings Ordinance, and the Renewable Energies Heat Act. This Act implements EU requirements for buildings and resolutions from the 2018 Housing Summit,[86] as well as the EU requirements for energy performance in buildings and the corresponding goals of the Federal Government's coalition agreement.[87] The GEG aims to ensure the most economical use of energy in buildings and to increase the use of renewable energies for heating, cooling, and electricity generation in buildings, as per §1 (1) GEG. In line with economic efficiency principles, the Act contributes to achieving the Federal Government's energy and climate policy objectives, increasing the share of renewable energies in final energy consumption for heating and cooling, and fostering sustainable energy supply development in the interest of climate protection, fossil resource conservation, and reduced dependence on energy imports, as stated in §1 (2) GEG. However, this law did not introduce any further tightening of the already stringent energy requirements set by the Energy Savings Ordinance.[88]

Furthermore, the Building Energy Act (Gebäudeenergiegesetz—GEG) does not create an entitlement for tenants to energy-efficient housing. Instead, it establishes a legal framework for the energetic modernisation of the existing building stock and sets legal requirements for the energy efficiency of new buildings, as specified in §47 (1) GEG. Neither the preceding Energy Savings Act nor the GEG grants tenants a right to demand energy modernisation measures from landlords.[89] However, the new GEG imposes insulation-related obligations on owners of both residential and non-residential buildings. Specifically, if a building is intended to be heated to an indoor temperature of at least 19 degrees Celsius for a minimum of four months per year, the upper floor ceilings must be

---

[84] Möbert 2021, p. 27; Stede/Schütze/Wietschel 2020, p. 775.

[85] Ibid, p. 775f.

[86] The Housing Summit took place on 21st of September 2018 at the Federal Chancellery. As a result, a paper "Joint Housing Initiative of the Federal Government, the Länder and the Municipalities" has been launched, more under: https://www.bmi.bund.de/SharedDocs/downloads/DE/veroeffentlichungen/2018/ergebnisse-wohngipfel.pdf;jsessionid=71CCDF-F3A2A5486966DD063F0E75BEF6.1_cid322?__blob=publicationFile%26v=7, accessed on: 22.10.2022.

[87] Manthey 2021, p. 147.

[88] Kaßler 2020, p. 58; Manthey 2021, p. 147.

[89] BGH NJW 2008, 142; Lindner 2019, Rn. 733.

insulated to meet the minimum thermal insulation requirements outlined in DIN 4108-2: 2013-02, ensuring that the heat transfer coefficient does not exceed 0.24 watts per square metre per Kelvin. This requirement is also considered satisfied if the roof is insulated in accordance with the same DIN standard (§47 (1) GEG).

However, there are exceptions to this legal requirement. Firstly, the obligations for existing buildings under Chap. 3 of the GEG are not applicable if compliance conflicts with other public law provisions related to structural safety, fire and noise protection, or labour and health protection, as stipulated in §46 (2) GEG. Additionally, in the case of a residential building with no more than two flats, one of which was owner-occupied as of 1 February 2002, the obligation under §47 (1) GEG applies to a new owner only if the property was transferred after this date. The deadline for compliance is two years from the initial transfer of ownership, as per §47 (3) GEG. Moreover, the aforementioned insulation requirements do not apply if the retrofitting expenses cannot be recouped within a reasonable timeframe through the resultant savings, as per §47 (4) GEG.

The term "reasonable period of time", as used in §47 (4) GEG, is an indeterminate legal concept. As previously mentioned, like any indeterminate legal concept, it requires definition by the executive and is fully subject to judicial review.[90] Currently, there is no established legal practice defining what constitutes a "reasonable period of time" under this provision. This norm has been critiqued by practitioners, who argue that it is rendered obsolete due to the general requirement for economic efficiency. The lack of specificity regarding what constitutes a "reasonable" period, particularly in relation to the service life of the insulation, opens the door for potential litigation and may render the requirement ineffective.[91] Despite the numerous exceptions outlined in §47 (1) GEG, the wording of this legal norm obligates building owners to undertake energy modernisation under certain conditions. However, this norm does not grant tenants the right to demand such modernisation.

In summary, low-income households in Germany have only a limited assurance of adequate housing. The rent for state-subsidised dwellings must be reasonable, with the definition of 'reasonable rent' and the criteria for its determination being a matter for the *Länder*. Social courts are tasked with ensuring that the state is not encumbered with unreasonable costs. A subsidised dwelling must fall within the lower rent price segment within its spatial district to be considered reasonable. However, the energy efficiency of dwellings is not a factor in determining reasonable costs. Moreover, low-income tenants do not have the right to demand energy-efficient modernisation of their dwellings in Germany.

---

[90] Schönenbroicher in: Mann/Sennekamp/Uechtritz 2019, VwVfG §40 Rn. 83f.

[91] Verband des Zimmerer- und Holzbaugewerbes Baden-Württemberg 2019, p. 5.

5.2.2.3 Legal Barriers Related to the Decarbonisation of Rented Dwellings in Germany

The next comparative research question addresses potential legal barriers to enhancing energy efficiency in rented dwellings under current legislation. Identifying these barriers will facilitate a comparison between the German and Swedish legal frameworks. This analysis will help assess whether there are transferable legal solutions to the landlord-tenant dilemma in Sweden, and if not, whether such solutions could be developed within the existing legal context.

Energy modernisation presents a delicate balance between various public and private interests, particularly in tight rental markets. Often, there are conflicting interests that pose challenges in achieving a universally acceptable or satisfactory resolution. On one hand, energy modernisation aligns with the public interest in climate protection. On the other hand, property owners have a vested interest in updating their properties to meet modern housing expectations and enhancing economic profitability. The preferences of solvent tenants may vary, with some favouring energy modernised dwellings for their increased housing value, despite potentially higher costs.[92] Conversely, low-income tenants often prioritise affordability over modernisation. Finding a middle ground that reconciles these diverse interests could alleviate, if not fully resolve, the tensions inherent in the landlord-tenant dilemma regarding energy efficiency upgrades.

Historical Background

Energetic modernisation, crucial for the decarbonisation of dwellings, is governed by various areas of law in Germany, with building-related regulations primarily shaped by federal law. At the European level, Art. 194 TFEU is the central legal norm in primary law that explicitly advocates the promotion of energy efficiency and savings as objectives of European energy policy,[93] as analysed in Chap. 4. The only directive on the energy efficiency of buildings is Directive 2010/31/EU of the European Parliament and of the Council, dated 19 May 2010, which focuses on the energy performance of buildings (recast). This directive significantly influences the formulation of energy efficiency regulations in the national tenancy laws of member states.[94] However, it does not set binding standards of requirements but instead mandates member states to establish these standards.[95]

Over the past decade, the German legislator has repeatedly responded to the challenges of the energy transition, adapting legal provisions accordingly. Since 2010, German tenancy law has undergone four reforms, which are briefly outlined here. In 2010, the Federal Government released an energy concept for an environmentally

[92] Gsell/Siegmund 2019, p. 489 f.

[93] BBSR 2016, p. 24.

[94] BBSR 2016, p. 25.

[95] Jesse 2016, p. 170.

friendly, reliable, and affordable energy supply.[96] This energy concept was further developed within the framework of the energy transition in spring 2011.[97] Consequently, the legislator was tasked with reforming existing tenancy law provisions, which had become inadequate for rented housing in terms of energy efficiency and climate protection.[98] As a result, the Act on the Energetic Modernisation of Rented Housing and on the Simplified Enforcement of Eviction Orders[99] (Mietrechtsänderungsgesetz—MietRÄndG) was enacted in 2013.

For the first time, the term energetic modernisation was legally defined (§555b No.1 BGB),[100] which refers to structural changes that result in sustainable final energy savings in relation to the rented property. The core of this reform was to simplify the enforcement of energetic modernisation measures by limiting the tenant's general right to reduce rent during renovation measures. Specifically, tenants can reduce rent in the case of energetic modernisation only after three months have elapsed, as stated in §536 (1a) BGB. Although landlords are still required to announce energy-related renovation measures, the requirements for this announcement were lowered (§555c (1) BGB).[101] Furthermore, a general obligation for tenants to tolerate modernisation measures in the interest of energy savings and climate protection was introduced (§555d BGB).[102]

In 2015, the German legislator addressed the issue of rising rents in re-let dwellings with the Act to Control Rent Increases in Tight Housing Markets and to Strengthen the Order Principle in Housing Brokerage (Mietrechtsnovellierungsgesetz—MietNovG).[103] New regulations in §§556d et seq. BGB were introduced, allowing rents to be increased to a maximum of 10% above the local comparable rent when re-letting existing dwellings. Additionally, state governments were empowered to designate, for up to five

[96] Bundestag, Drucksache 17/3049.

[97] Bundestag, Drucksache 17/10485, p. 1.

[98] Ibid.

[99] Gesetz über die energetische Modernisierung von vermietetem Wohnraum und über die vereinfachte Durchsetzung von Räumungstiteln vom 11. März 2013, BGBl. II p. 434f.

[100] BeckOGK/ *H. Schmidt,* BGB §535 Rn. 16; Gesetz über die energetische Modernisierung von vermietetem Wohnraum und über die vereinfachte Durchsetzung von Räumungstiteln vom 11. März 2013, BGBl. II p. 435; Blank/Börstinghaus in: Blank/Börstinghaus 2020, Miete, §555b BGB, Rn. 1.

[101] BeckOGK/ *H. Schmidt,* BGB §535 Rn. 16; Gesetz über die energetische Modernisierung von vermietetem Wohnraum und über die vereinfachte Durchsetzung von Räumungstiteln vom 11. März 2013, BGBl. II pp. 435.

[102] Gesetz über die energetische Modernisierung von vermietetem Wohnraum und über die vereinfachte Durchsetzung von Räumungstiteln vom 11. März 2013, BGBl. II p. 435.

[103] Gesetz zur Dämpfung des Mietanstiegs auf angespannten Wohnungsmärkten und zur Stärkung des Bestellprinzips bei der Wohnungsvermittlung vom 21. April 2015, BGBl. I, p. 610ff; Bundestag, Drucksache 18/3121.

years until the end of 2020, apply, using statutory instruments.[104] However, dwellings first used and rented after 1 October 2014, and those rented for the first time after extensive modernisation, were excluded from the scope of §556d BGB, as per §556f BGB.[105] Furthermore, a legal right for tenants to obtain information from landlords about rent-forming factors was established, provided these factors were not generally accessible and the landlord could readily provide such information (§556g (3) BGB).[106] Other amendments to the tenancy legislation in 2015, not pertinent to the landlord-tenant dilemma, will not be discussed here.

The rent control measures introduced by the MietNovG did not achieve the desired effect. Rents continued to rise significantly, even in areas designated as tight housing markets under §556d (2) BGB. This was partly due to informational barriers faced by tenants. As the weaker party, they often lacked the ability to assess, without significant effort, whether the rent demanded was acceptable, especially under exceptional circumstances, hindering their ability to legally challenge increased rents.[107] In response, the legislator enacted the Act Supplementing the Regulations on the Permissible Amount of Rent at the Beginning of the Tenancy and Adjusting the Regulations on the Modernisation of the Rented Property (Mietrechtsanpassunggesetz—MietAnpG), which came into force on 1 January 2019.[108] This act aimed to establish a fair balance between landlord and tenant interests, as the previous regulatory situation increasingly burdened tenants, compelling them to exercise their special right of termination following a modernisation notice under §555e BGB. A key factor behind these burdens was a significant increase in energetic modernisation, partly driven by the Federal Government's "Energy Efficiency Strategy for Buildings,"[109] which aimed to combine energy savings with the use of renewable energies.[110]

The Mietrechtsanpassunggesetz (MietAnpG) introduced key changes concerning energy modernisation. One significant amendment was the reduction of the modernisation levy rate in §559 (1) BGB from 11% to 8% for landlords who undertook measures for energetic modernisation, sustainable water consumption reduction, utility value enhancement of the rental dwelling, and/or permanent improvement of general living conditions. Additionally, §559 (3a) BGB introduced a cap on modernisation to safeguard

---

[104] According to the latest amendment, a statutory order regulating rent restrictions shall expire at the latest by the end of 31 December 2025, §556d (2) S.4 BGB.

[105] Bundestag, Drucksache 18/3121, S.1f.

[106] Ibid.

[107] Bundestag, Drucksache 19/4672, S. 12; Börstinghaus 2018, p. 487.

[108] Gesetz zur Ergänzung der Regelungen über die zulässige Miethöhe bei Mietbeginn und zur Anpassung der Regelungen über die Modernisierung der Mietsache vom 18. Dezember 2018, BGBl. I p. 2648ff.

[109] BMWi 2015.

[110] Börstinghaus 2018, p. 490.

tenants from affordability issues due to substantial rent increases following modernisation.[111] A simplified procedure for rent increases was established in §559c BGB for modernisation measures costing no more than €10,000 per flat.[112]

To protect tenants from so-called "out-modernisation" (*Herausmodernisierung*), the legislator introduced legal presumptions for breaches of duty in connection with structural changes that were incorporated in §559d BGB. This provision applies if a landlord uses the announcement of constructional changes to pressure the tenant into terminating the tenancy.[113] Furthermore, tenant protection was further reinforced by amending §6 WiStG (Act on the Further Simplification of Commercial Criminal Law)[114]. Under this amendment, it constitutes an administrative offence, punishable by a fine of up to €100,000 (§6 (2) WiStG), if a landlord undertakes or commissions structural modifications with the intention of inducing a tenant to terminate their tenancy, leading to significant, objectively unnecessary burdens for the tenant.[115]

In 2021, the legislator introduced the latest amendments to German tenancy law by enacting the Act on the Reform of Rent Index Law (Mietspiegelreformgesetz—MsRG).[116] This Act aims to strengthen the legal certainty of the rent indices. The comparative rent system is one of the guiding principles of social tenancy law as it ensures legal certainty and a fair balance between the interests of landlords and tenants. Recently, the validity of these rent indices has been increasingly challenged in court proceedings, particularly regarding their adherence to recognised scientific principles.[117]

Tenancy Law Regulations on Energy Modernisation

As previously noted, the German legislator has established two distinct procedures for rent increases following modernisation measures: a 'usual' procedure and a simplified procedure. Both procedures, particularly in the context of energy modernisation as outlined in §555b No. 1 BGB, will be scrutinised in detail in the following sections.

---

[111] Bundestag, Drucksache 19/4672, p. 13.

[112] Ibid.

[113] Ibid.

[114] Gesetz zur weiteren Vereinfachung des Wirtschaftsstrafrechts (Wirtschaftsstrafrecht 1954) vom 9.07.1954, BGBl. I, p. 175ff.

[115] Gesetz zur Ergänzung der Regelungen über die zulässige Miethöhe bei Mietbeginn und zur Anpassung der Regelungen über die Modernisierung der Mietsache vom 18. Dezember 2018, BGBl. I p. 2650.

[116] Gesetz zur Reform des Mietspiegelrechts (Mietspiegelreformgesetz—MsRG) vom 10.08.2021, BGBl. I, p. 3515ff.

[117] Bundestag, Drucksache 19/26918, p. 1; AG Mainz Urt. v. 08.05.2013—86 C 323/11; BGH Urt. v. 06.11.2013 – VIII ZR 346/12.

The "usual" Procedure

In the 'usual' procedure for rent increase following modernisation, the landlord is required to inform the tenant about the planned modernisation of the residential property in text form. This means communication methods like email or fax are sufficient, as per §126b BGB.[118] However, notices placed in communal areas like hallways or notice boards do not meet these requirements. If there are multiple tenants, the modernisation notice must be addressed to all of them.[119]

When discussing energy modernisation in this context, it's important to highlight that, according to §555b No. 1 BGB, the modernisation measure must be related to the rented property, encompassing both the individual dwelling and the residential building. Energy-saving measures that do not directly relate to the rented property fall under §555b No. 2 BGB. For instance, installing a photovoltaic system on the roof of a rented building is categorised under §555b No. 2 BGB, as the generated electricity benefits the general electricity grid rather than the building itself.[120] Additionally, energy modernisation under §555b No. 1 BGB requires the sustainability of savings, meaning they must be of a permanent nature;[121] temporary savings effects are insufficient.[122]

The requirements for the content of the modernisation notice are governed by §555c BGB. The landlord must provide substantial information to the tenant about the nature, expected scope, commencement, and duration of the modernisation measures. This includes details about the anticipated rent increase due to modernisation and future utility costs.[123] The notice enables the tenant to assess their legal obligation to tolerate the modernisation and to decide whether to exercise their special right of termination under §555e BGB. This notice should be given in text form at least three months before the commencement of the modernisation measures, as specified in §555c (1) BGB. While the landlord is expected to describe the modernisation measures concretely, avoiding vague statements (like "energy savings through a new heating system"), they are permitted to refer to generally recognised lump-sum values, particularly concerning the energy efficiency of building components, as per §555c (3) BGB.[124]

---

118 Derleder 2013, p. 446; Blank/Börstinghaus in: Blank/Börstinghaus 2020, Miete, §555c BGB, Rn 5; Schultz in: Bub/Trier 2019, Rn. 1476.

119 Blank/Börstinghaus in: Blank/Börstinghaus 2020, Miete, §555c BGB, Rn 5;

120 Blank/Börstinghaus in: Blank/Börstinghaus 2020, Miete, §555b BGB Rn. 20; Lutz in: Hannemann/Wiegner 2019, §23 Rn. 224.

121 Blank/Börstinghaus in: Blank/Börstinghaus 2020, Miete, §555b BGB Rn. 21.

122 Blank/Börstinghaus in: Blank/Börstinghaus 2020, Miete, §555b BGB Rn. 21; BGH NJW 2002, 2036; Eisenschmid in: Schmidt-Futterer 2022, Mietrecht, §555b BGB Rn. 26; Lindner 2019, Rn. 737.

123 BeckOK MietR/*Müller*, §555c Rn.23.

124 Derleder 2013, p. 447.

Tenant's Obligation to Tolerate the Energy Modernisation

Under §555d (1) BGB, tenants are generally required to tolerate modernisation measures. However, there is an exception concerning a hardship clause as per §555d (2) BGB. The evaluation of hardship under this clause necessitates a balancing of the interests of the tenants, other occupants in the building, and the broader interests of energy-saving and climate protection, reflecting public interest. Notably, the legislator distinguishes between personal and economic reasons for hardship. Consequently, this regulation does not extend to anticipated increases in rent and utility costs due to modernisation. The assessment of hardship in relation to increased utility costs following energy modernisation is governed by §§559 (4) and (5) BGB. According to §555d (3) BGB, if a tenant believes that a modernisation measure causes them hardship, they must inform the landlord of this in text form within one month of receiving the modernisation notice. It is important to note that this one-month period does not commence until the modernisation notice fulfils the requirements set out in §555c BGB.[125]

Calculation of the Modernisation Rent Increase

According to §559 (1) BGB, the landlord is permitted to increase the annual rent by 8% of the costs incurred for the flat, provided they have undertaken energy modernisation measures.[126] The annual rent is understood as the total rent paid in the preceding year or twelve times the amount of rent paid in the last month before the modernisation. This interpretation is clarified in the explanatory statement of the Tenancy Law Reform Act concerning Section 559 BGB (Mietrechtsformgesetz zu §559 BGB).[127] The prerequisite for this rent increase is that the modernisation measures are carried out in the landlord's name and at their expense. If modernisation is conducted by a third party at their expense, §559 BGB does not apply, and it is instead limited to §558 BGB, provided that the modernisation results in improved equipment or condition.[128]

Additionally, costs associated with maintenance measures are not included in the incurred costs for modernisation. These costs, if necessary, are to be determined by estimation as per §559 (2) BGB. The landlord bears the responsibility of proving the incurred costs, including those related to maintenance expense.[129] According to §559a BGB, any third-party funds must be deducted from a modernisation rent increase. Moreover, financing costs, such as the cost of raising capital, and rent reductions or loss of rent due to construction measures, do not qualify as "incurred costs".[130]

[125] Derleder 2013, p. 447;

[126] Fleindl 2016. p. 67; Artz in: MüKoBGB, §559, Rn. 16.

[127] BT Drucksache 14/4553, p. 58; Arzt in: MüKoBGB, §559 Rn. 14; Lutz in: Hannemann/Wiegner 2019, §23 Rn. 237.

[128] Fleindl 2016. p. 67.

[129] Fleindl 2016, p. 73; BGH NZM 2009, 150 Rn. 19.

[130] Artz in: MüKoBGB, §559 Rn. 17; OLG Hamburg NJW 1981, 2820.

Importantly, §559 (2) s.1 BGB provides potential solutions to the landlord-tenant dilemma by prohibiting rent increases if the energy modernisation is financed through third-party costs. In cases where energy modernisation measures are funded by the landlord, the legislator has implemented a cap on rent increases through §559 (3a) BGB. The rent cannot be increased by more than €3 per m^2 of living space per month within six years following modernisation. If the monthly rent prior to the increase is below €7 per m^2 of living space, the increase is capped at €2 per m^2.[131] This cap ensures predictability for the tenant regarding post-modernisation rent increases. However, it simultaneously diminishes the financial incentive for landlords to undertake such measures. This regulation does not yet apply to rent increases up to the local comparative rent under §558 BGB and rent increases due to changes in utility costs under §560 BGB.[132] As these are independent of energy modernisation measures, they will not be further discussed in this thesis.

The "simplified" Procedure

Modernisation Notice

If a landlord decides to make use of the simplified procedure, he must notify the tenant of this decision in the modernisation notice as required by §559c (3) s. 2 BGB. When employing this approach, the landlord is allowed to forego providing details of the expected utility costs in the modernisation notice, which would typically be mandatory under §555c (1) s. 2 No. 3 BGB, as per §559c (5) No. 2 BGB. This is particularly beneficial in cases of energy modernisation, where calculating the projected utility cost savings and, consequently, the anticipated future utility costs can be a complex and time-consuming task.[133]

Furthermore, similar to the requirements under §559b BGB, the landlord must provide a clear explanation of the modernisation measure and the allocation of costs in the modernisation notice. If the landlord chooses to limit the claimed costs to take advantage of the simplified procedure, they must also present a corresponding calculation. However, under the flat-rate method, providing information on the share of maintenance costs is not necessary.[134]

Tenant's Obligation to Tolerate the Energy Modernisation

When a landlord selects the simplified procedure for increasing rent, it has a specific implication for the tenant: they cannot claim a hardship case in this scenario. This is because, under §559c (1) s. 3 BGB, the provisions of §559 (4) BGB, which typically

[131] Bundestag, Drucksache 19/4672, p. 31; Lutz in: Hannemann/Wiegner 2019, §23 Rn. 242.

[132] Bundestag, Drucksache 19/4672, p. 31.

[133] Artz/Börstinghaus 2019, p. 20; Artz in: MüKoBGB, §559c Rn. 8.

[134] Artz in: MüKoBGB, §559c Rn. 8.

allow for hardship considerations, do not apply to the simplified procedure. Additionally, apart from this distinction, there are no other significant differences concerning the tenant's obligation to tolerate the rent increase under the simplified procedure as compared to the standard procedure.

Calculation of the Modernisation Rent Increase

The simplified procedure for rent increases, which offers less protection for tenants, is applicable only for smaller modernisation measures, with an investment limit of up to €10,000 per flat. Furthermore, a mandatory deduction of 30% of the incurred costs for maintenance must be applied, as stipulated in §559c (1) s. 1 h.-s. 2 BGB. This legal provision refers to the modernisation costs claimed for the flat, allowing the landlord the option to use the simplified procedure even for larger measures by limiting the scope of the costs passed on to the tenant to €10,000. However, this approach is generally not economically viable for measures where the actual investment costs significantly exceed €10,000 per flat.[135]

The landlord cannot claim the absence of maintenance work in this context. In certain scenarios, it is not possible to deduct third-party funds when calculating a rent increase due to modernisation. For instance, if the modernisation is financed through public credit instruments, the loan amount does not need to be deducted as a third-party fund. Additionally, the landlord is not required to deduct the interest advantage of a subsidised or interest-free loan as per §559a (2) BGB, because, according to §559c (1) s. 3 BGB, §559a (2) s. 1 to 3 do not apply to the simplified procedure. However, if the landlord has received subsidies from public budget funds or if third parties have borne the costs of modernisation, these must be deducted from the incurred costs in accordance with §§559a (1), (2) s.4 BGB. The rationale for these regulations is that public funds should ultimately benefit the tenant.[136]

Moreover, as per §559c (2) BGB, the costs incurred before the deduction of maintenance expenses from all modernisation rent increases in the last five years prior to the receipt of the modernisation increase notice are counted towards the €10,000 cap. If this maximum amount is exceeded, the landlord must implement the rent increase in accordance with the standard procedure outlined in §559 BGB.

Land Law Barriers

The concept of rent control, or "*Mietpreisbremse*", was recently incorporated into German tenancy law in 2015 with the amendments to the BGB previously discussed. However, the notion of rent control has been a subject of political discussion since 1916, initially emerging as a response to pressures during the First World War. In recent years,

[135] Artz in: MüKoBGB, §559c Rn. 2; Lutz in: Hannemann/Wiegner 2019, §23 Rn. 278ff.

[136] Artz/Börstinghaus 2019, p. 20.

the rapid increase in rents in major German cities and university towns has become a central concern in housing policy, necessitating state intervention for social reasons.[137]

Due to the strained conditions in Berlin's rental housing market, the state of Berlin decided to intervene. The primary cause of tension in the rental market is the continuously increasing demand for housing, which has not been matched by a corresponding rise in housing supply through new construction. Tenants, particularly in the absence of alternatives, have often been compelled to accept rising rents, especially after modernisation. This situation poses a significant threat of displacement for low-income tenants from their social environment.[138]

As a measure to address these challenges, Berlin introduced the Law on Rent Limitation in Housing in Berlin (MietenWoG Bln) in the winter of 2020. The law aimed to cap rent increases in Berlin for five years. Under this law, any rent exceeding the amount effectively agreed upon on 18 June 2019 (the reference date) was deemed void (as per §3 (1) MietenWoG Bln). The law also included special provisions for modernisation-related rent increases (see §7 MietenWoG Bln).[139]

In 2021, the BVerfG declared the MietenWoG Bln incompatible with the GG and, therefore, void.[140] The concurrent legislation in tenancy law does not permit the *Länder* to enact their regulations, as the federal government has exhaustively utilised its legislative competence. Moreover, the regulations of this law contradicted the BGB regulations.[141] The BVerfG's decision to annul the Berlin rent control measures served as a clear signal to other *Länder* against intervening in rent increases at the state level. However, this decision did not address the issue of the lack of incentives for energy modernisation due to the existing federal caps on rent increases.

5.2.2.4 Critical Evaluation of the Legal Situation

At this point, a critical evaluation of the current legal framework is essential to address the third research question of this study: whether the legal barriers impeding the energy modernisation of rented dwellings occupied by low-income households can be surmounted through reforms in tenancy law. This evaluation will involve examining the effectiveness of existing legal provisions, their impact on low-income tenants, and the potential for legal amendments to facilitate more accessible energy modernisation in rental housing.

[137] Herrlein/Tuschl 2020, p. 218.

[138] Abgeordnetenhaus Berlin. Drucksache 18/2347, p. 1.

[139] Karras/Novikova 2021, p. 28.

[140] BVerfG, NJW 2021, 1377.

[141] Ibid, 1378f.

Landlord-tenant Dilemma from the Perspective of German Tenancy Law

After a thorough analysis of the legal regulations on energy modernisation measures, it is evident that a landlord-tenant dilemma emerges when the tenant benefits from the energy modernisation of the building, without corresponding advantages accruing to the landlord.[142] German tenancy law incorporates a significant social component, offering special protection to the tenant. This is a manifestation of the state's obligation to create and guarantee decent living conditions, as enshrined in Art. 1 (1) of the GG in conjunction with the welfare state principle under Art. 20 (1) of the GG.[143] This social component exerts a strong influence on the housing market, which operates beyond the mere principles of a free economy. The numerous hardship clauses, such as the obligation to tolerate hardship in §555d (2) s.1 BGB and the hardship clause related to rent increase in §559 (4) BGB, prioritise social protection over climate protection.[144]

The landlord faces several risks, including the possibility of the tenant exercising their special right of termination under §561 (1) 1 BGB following a modernisation notice, potentially leading to vacancy and income loss. Additionally, there is the risk that the tenant may claim hardship, preventing any rent increase. Even in cases where rent increase is permitted, the landlord's options are significantly constrained.

Consequently, landlords and tenants are not on equal economic footing when it comes to post-modernisation scenarios. Tenants benefit from lower heating costs and improved living standards, but even if the landlord increases the rent, it often does not equate to the market value of the modernisation undertaken. Therefore, under the current legal framework, landlords lack incentive to invest in energy modernisation. This situation underscores the existence of a landlord-tenant dilemma in Germany, where the legal framework inadvertently discourages landlords from undertaking energy-efficient upgrades.

Legal Barriers on the Way to Energy Modernisation of Rented Building Stock

Prior to the enactment of the Mietrechtsanpassungsgesetz (MietAnpG), the barriers contributing to the landlord-tenant dilemma were already significant. However, the introduction of the MietAnpG only served to intensify these challenges. The legislation has been subject to considerable criticism in academic and practitioner circles. It was often described as a symbolic law, addressing immediate conflicts and day-to-day situations without effecting substantial change.[145] Although concerns were raised about the constitutionality of the MietAnpG,[146] hese were not upheld by the Federal Constitutional Court. Nonetheless, the reform attracted strong criticism for various reasons.

---

[142] Hallof 2013, p. 26f.

[143] Ruhland/Kreibich/Herud 2010, p. 3.

[144] Derleder 2013, p. 443.

[145] Arzt/Börtinghaus 2019, p. 25.

[146] Abramenko 2018, p. 36; Mietverein Berlin, 2019.

One major point of contention was the reduction of the permissible rent increase following modernisation from 11% to 8%. Critics argued that this reduction would have detrimental effects. Landlords may be disinclined to make investments that do not align with the reduced rent increase, as this could threaten the profitability of their property. The modernisation levy is typically intended to cover additional maintenance costs, amortisation, and investment expenses incurred due to modernisation.[147] However, even prior to this reduction (when a rent increase of up to 11% was possible), landlords often could not fully utilise the available scope for rent increases for various reasons.

The market response did not adequately compensate for the improvements in comfort resulting from modernisation. Additionally, the energy cost savings achievable in the short term did not match up with the legal possibilities for rent increases. This situation was already present when the rent increase cap was at 11%. Now, with the reduction to a cap of 8%, there has been a further decline in landlords' willingness to invest in energy modernisation. This reduced incentive is a significant concern, as it potentially hinders efforts to improve energy efficiency in rental properties.[148]

The impact of the modernisation levy varies significantly depending on the housing market. In urban areas like Berlin, where there is high demand for housing, tenants are more willing to pay increased rents. This demand also extends to apartments with higher rents following energetic modernisation,[149] creating an added barrier for low-income households who cannot afford energy-efficient housing. Conversely, in these dynamic markets, landlords may not be deterred from delaying energy modernisation due to the low risk of vacancy.[150] However, in rural markets with lower demand, such as the district of Spree-Neiße, landlords face the risk that increased rents post-modernisation may not be enforceable, heightening investment risks.[151] Yet, energy modernisation could enhance the attractiveness of these properties and mitigate the risk of vacancy.[152]

The German legislator acknowledged the risks associated with increasing the modernisation levy in the preparation of the MietAnpG. The explanatory memorandum to the draft law highlighted concerns that landlords might forego extensive modernisation measures, including those related to energy efficiency.[153] Nonetheless, the legislator considered that the MietAnpG balanced the need for tenant protection with broader social goals related to climate change and demographic challenges. The acceptance of potentially unachieved energy-saving effects was justified on socio-political grounds.[154]

---

[147] Henger et al. 2021, p. 17.

[148] Bundestag, Drucksache 19/4672, p. 5 f.; Gsell/Siegmund 2019, p. 499.

[149] Henger et al. 2021, p. 10.

[150] Ibid, p. 11.

[151] Ibid, p. 10.

[152] Ibid, p. 11.

[153] Bundestag, Drucksache 19/4672, p. 24.

[154] Bundestag, Drucksache 19/4672, p. 24.

In this context, achieving complete, negotiated solutions for energy-related building renovations under market conditions poses challenges. Effective tenancy law should provide mutual benefits to both parties from energy modernisation, ensuring that any agreement on such renovations reflects market conditions at the time of modernisation and throughout the lease term.[155]

From an economic analysis and game theory perspective, under the current legal framework, it is more likely that both parties—landlords and tenants—will defect (Figure 1, Scenario IV) to avoid harmful risks. Landlords lack incentives to undertake energy modernisation measures exceeding €10,000 per dwelling due to prolonged cost recovery periods and the unpredictability of tenant behaviour. This makes mid-term profitability uncertain.

Given the pressing need in densely populated areas like Berlin to provide affordable rental housing and considering Germany's climate protection goals, resolving the landlord-tenant dilemma through tenancy law reform alone seems infeasible without compromising on energy modernisation. Therefore, a solution beyond the realm of tenancy law is necessary.

5.3 Swedish Context

5.3.1 Non-legal Context and Institutional Framework

The Kingdom of Sweden (*Konungariket Sverige*), located in Northern Europe, is a Scandinavian country known for its extensive forests, which cover nearly half of its territory. A significant portion of Sweden, particularly in the north, lies on or above the Arctic Circle, making it largely uninhabitable or sparsely populated.[156] As of January 2022, Sweden's population stands at approximately 10.1 million inhabitants.[157] The Swedish form of government is a monarchy, and it is a unitary state.[158] The monarch is the representative head of state, yet his role is weak as Sweden's governmental form is a constitutional parliamentary democracy.[159]

The Swedish political and administrative system is characterised by its decentralisation and fragmentation[160] across three levels: national, regional, and local.[151] Sweden

[155] Hallof 2013, p. 197.

[156] Ortwein 2003, pp. 406f.

[157] Population of Sweden (2022), in: <https://www.worldometers.info/world-population/sweden-population/>, accessed on: 26.01.2021.

[158] Dotevall in: Smits 2006, p. 700.

[159] Ortwein 2003, p. 408.

[160] Pierre/Jochen/Jahn 2016, p. 2.

[161] Government offices of Sweden 2020, p. 4.

consists of 21 counties (*län*), divided into 290 municipalities (*kommuner*). Each county has a government authority—the County Administrative Board (*länstyrelse*) and a governor (*landshövding*). These positions are non-elective, and they are a part of the state's local administration.[162] At the municipal level, a locally elected municipal assembly (*kommunfullmäktige*) leads each municipality, appointing a municipal council (*kommunstyrelse*) as its executive head.[163]

The County Administrative Boards act as a liaison among citizens, municipal authorities, the government, parliament, and central authorities. They are tasked with implementing and administering national political goals within their respective counties. The municipalities have the competencies to execute and provide a significant share of all public tasks and services, including environmental protection.[164] Local and regional governments are autonomous, which is guaranteed constitutionally.[165]

Municipalities, counties, and regions in Sweden operate independently without any hierarchical ranking, each with distinct areas of responsibility.[166] Local and regional governments predominantly self-finance through local and county council taxes.[167] While the constitution does not explicitly outline municipal tasks, there is a longstanding tradition of political autonomy at the municipal level in Sweden.[168]

Sweden has experienced rapid urbanisation in recent years, with about 40% of the population residing in metropolitan areas. Over two million people live in Stockholm, representing approximately 22% of the national population. Gothenburg and Malmö are other significant metropolitan areas. This trend towards urbanisation is expected to persist, with metropolitan areas projected to expand further. However, this growth contrasts with the declining population trend in smaller municipalities with fewer than 25,000 inhabitants, as people increasingly relocate to larger municipalities.[169]

Despite the relatively low population and moderate economic growth, Sweden has been a forerunner in environmental policy as it was on the political agenda in the 1970 s.[170] Municipalities have played an important role in Swedish energy planning for over three decades, with the Swedish legislator mandating the development of energy plans since 1977.[171] Historically, Sweden's industrial economy was highly

---

162 Carlson 2012, p. 32; SEPA 2017, p. 10.

163 Buhr/ Frankenberger 2016, p. 1.

164 SEPA 2017, p. 10.

165 Pierre/Jochen/Jahn 2016, p. 2; SEPA 2017, p. 10.

166 Carlson 2012, p. 32f.

167 Government offices of Sweden 2020, p. 4.

168 Buhr/ Frankenberger 2016, p. 1.

169 Bååth 2014<https://www.tenlaw.uni-bremen.de/Brochures/SwedenBrochure_09052014.pdf>, p. 11.

170 OECD 2017; Femenias/ Lindén 2010, p. 1.

171 Nilsson/ Mårtensson 2003, p. 180.

energy-intensive, predominantly reliant on oil products (constituting around 75% of energy supply). The oil crisis of the 1970 s hit Sweden hard, prompting the government to launch an ambitious campaign aimed at reducing energy consumption. From 1975 to 1985, Sweden successfully reduced the oil share of its energy supply from under 70% to below 40%, while simultaneously increasing the share of nuclear power energy from zero to around 15%. Additionally, the development of district heating systems, primarily managed by municipalities, and the use of biofuels significantly enhanced energy efficiency during this period. From 1979 to 1985, municipalities were required to develop complementary municipal oil reduction plans, transitioning local oil-based energy systems to alternative energy sources (including nuclear power) and more energy-efficient technologies. However, since the mid-1980 s, preparations for the phase-out of nuclear power commenced.[172]

In terms of greenhouse gas emissions, Sweden achieved a notable reduction between 1970 and 1990. In 1988, the Riksdag resolved to stabilise carbon dioxide emissions at the 1988 level, expanding this target three years later to encompass all greenhouse gases across various economic sectors. Furthermore, Sweden was a pioneer in introducing a carbon tax in 1990–1991 as part of a comprehensive tax reform that revised existing environmental and energy taxes. This reform imposed financial responsibility on polluters, incentivising emission reduction.[173] Over three decades, the carbon dioxide tax increased almost fivefold, from SEK 0.25 per kg of carbon dioxide in 1991 to SEK 1.20 per kg in 2021.[174] An interesting fact is that Sweden had earlier introduced an energy tax on petrol in 1924 and on diesel in 1937. In the 1950 s, fuels used for heating and electricity generation were also taxed. However, the primary aim of the energy tax was fiscal, without a direct focus on steering energy usage towards higher efficiency or climate objectives.[175]

In the last decade of the 20th century, Sweden adopted its national climate strategy, aligning with the UNFCCC's goal to stabilise emissions in industrialised countries. Sweden committed to stabilising its carbon dioxide emissions from fossil fuels by 2000, relative to 1990 levels, and aimed for a subsequent reduction. In 1999, the *Riksdag* incorporated the objective to reduce climate impact into its environmental goals system. In the 2000s, the bill "Sweden's Climate Strategy" set new climate policy goals, including the target to limit CO_2 equivalent emissions to 4.5 tonnes per year per capita by 2050. Additionally, Sweden set an ambitious goal to achieve a fossil-fuel-independent vehicle fleet by 2030. By 2009, the country aimed for net-zero emissions by 2050, which was

[172] Swedish Climate Policy Council 2019, pp. 12–13; Nilsson/ Mårtensson 2003, p. 181.

[173] Swedish Climate Policy Council 2019, pp. 12–13.

[174] Naturvardsverket 2021, p. 12.

[175] Ibid, p. 13.

further advanced in 2017 with the new climate policy framework, setting the goal of net-zero greenhouse gas emissions by 2045.[176]

The Swedish economy and Swedish regulation thereof may be characterised as highly competitive and efficient. However, some sectors of the economy, e. g. the housing market, are criticised for being low efficient and untransparent.[177] Housing shortages, exacerbated by inadequate construction and demographic changes, are a significant concern.[178] Existing tax rules contribute to societal indebtedness and inequalities, particularly in the housing market. Rental regulations also pose barriers to efficient use of existing housing stock and mobility. Furthermore, inefficient land-use planning and low local-level incentives hamper housing supply, and a lack of competitiveness in construction leads to high costs.[179]

Sweden exemplifies good practices in environmental policy implementation. The government plans to introduce a state subsidy for property owners undertaking energy efficiency renovations in apartment buildings with rental dwellings, thereby improving the profitability of energy modernisation. This subsidy is intended for additional costs associated with such modernisation, with planned support amounting to 2.4 billion SEK in 2022 and 1 billion SEK in 2023.[180] Sweden was among the first countries to actively utilise green financing instruments. In 2019, green bonds accounted for 20% of the total bond market issued in Swedish krona. That year, the Swedish krona ranked fourth, after the EUR, USD, and CNY, in terms of all green bond emissions. Swedish municipalities actively participate in EU environmental protection and climate change initiatives,[181] with Gothenburg being the first to issue a green bond in 2013,[182] followed by Vasakronan, a property company, issuing the first green corporate bond globally.[183]

5.3.2 Legal Framework

5.3.2.1 General Overview

The Swedish legal system belongs to the Nordic legal family. Comparative law scholars often categorise the Nordic legal family as a branch within the Roman-Germanic legal family; however, this appears to be an oversimplification.[184] The prevailing opinion does

---

[176] Swedish Climate Policy Council 2019, p. 14.

[177] Pierre/Jochen/Jahn 2016, p. 8.

[178] European Commission, SWD (2019) 1026 final, p. 4.

[179] OECD Sweden 2019, p. 37.

[180] Naturvardsverket 2021, p. 25.

[181] Ferlin/ Sternbeck Fryxell 2020, pp. 6–7; Karras/ Novikova 2021, p. 4.

[182] Karras/ Novikova 2021, p. 4.

[183] Ferlin/ Sternbeck Fryxell 2020, p. 7.

[184] Bogdan 2010, p. 11.

not classify the Swedish legal system either as common law or to civil law system at all.[185] Instead, Swedish law is intermediate between continental law and the common law system.[186] Similar to the Roman law tradition, which has influenced the Swedish one, there is a categorisation of areas of law into public and private law.[187]

It is essential for Swedish lawyers that jurisprudence and legal practice remain aligned with common sense. The Swedish legal approach can be described as pragmatic, closely connected to real life, and with minimal emphasis on scientification.[188] Swedish law has been dramatically influenced by the jurisprudential movement known as Scandinavian Legal Realism, founded by the Swedish philosopher *Axel Hägerström* and the Danish philosopher and jurist *Alf Ross*.[189] The main aim of this movement was to eliminate any metaphysical influences on scientific thinking, in general, and legal thinking, in particular. This aims to secure a scientific understanding of law and legal science relevant to the lives of people within a state.[190]

It should be mentioned that Scandinavian Legal Realism, particularly the specific legal and philosophical streams (e.g., the Uppsala school), is based on complex philosophical ideas that are challenging to understand and convey to a lawyer without extensive prior experience in this matter.[191] Therefore, the influence of this philosophical stream on Swedish law is only described here in a concise and didactically simplified manner.

According to *Kischel*, there is only an indirect influence of legal realism on law today. First and foremost, it influences the persistent focus of Swedish jurisprudence on the law in force, its enforcement and its structures, and the education's emphasis on preparing future judges and lawyers. Social, political, philosophical, or economic aspects are not considered jurisprudential issues and must be researched by the respective disciplines. Furthermore, the will of the legislator, as expressed in the preparatory materials of the parliament, plays a crucial role in the Swedish doctrine of legal interpretation, whereas independent evaluations by jurists are secondary.[192] For the comparative analysis of this thesis, it means that the answers to the research questions, which have an economic or social background, should be researched in these disciplines and not in the legislation.

The fundamental structure, methods, way of legal thinking and a general understanding of law in the Swedish legal system is different to the German one and has been influenced by historical, social, and cultural development. However, EU membership has

[185] Carlson 2012, p. 36, Kischel 2015, p. 594f.; Strömholm 2013, p. 30ff.

[186] Dotevall in: Smits 2006, p. 699.

[187] Carlson 2012, p. 37.

[188] Kischel 2015, p. 608.

[189] Bjarup 2005, p. 1; Kischel, pp. 620f.

[190] Bjarup 2005, p. 1.

[191] Kischel 2015, p. 622.

[192] Ibid, pp. 625f.

required changes in the legal system and culture; for example, Swedish lawyers assumed a more international focus.[193]

Sweden, like Germany, has a dualistic system in the context of international law. This means that international law norms, such as international treaties, must be implemented into Swedish legislation to be applicable in this country. However, Swedish authorities are obliged to interpret the Swedish laws in accordance with international treaties ratified by Sweden. But in case of collusion between a national and an international legal norm, the national legal norm still prevails.[194]

Unlike in many other countries, the Swedish Constitution (*grundlagarna*) includes four fundamental laws: the Instrument of Government, the Act of Succession, the Freedom of the Press Act and the Fundamental Law on Freedom of Expression.[195] Additionally, there is the Riksdag Act, which regulates the work of the Riksdag (Parliament). This act, however, is not constitutional, despite the fact that special rules govern its amendment process.[196]

One of the unique features of Swedish jurists is the extremely cautious attitude of jurists towards judicial control of the legislature, resulting in a relatively low significance of constitutional law.[197] The role of fundamental rights in Sweden differs significantly from, for example, Germany. In Swedish constitutional law, fundamental rights do not play a crucial role.[198] This distinction will become clearer when the role of fundamental rights is explained later (Sub-chapter 3.2.2.1).

A critical aspect of the relatively insignificant role of constitutional law in Sweden can be attributed to its pronounced welfare state and its corresponding attitude towards state power. In this context, the Swedish state does not necessitate precise boundaries for its legislative and administrative actions. Rather than being viewed as a potential threat to citizens' fundamental rights and freedoms, the state is perceived as a guarantor of their well-being.[199] Consequently, restricting the legislature with overarching values, such as fundamental human rights, is seen as contradictory to the principles of legal realism. Additionally, independent evaluations within the framework of judicial decisions, as in the case of fundamental rights, are not typical in Sweden.[200]

The primary source of law in Sweden comprises laws and other regulations. These can be differentiated into constitutional acts, ordinary acts, ordinances, and statutory

---

193 Ortwein 2003, p. 445.

194 Grundberg 2016, p. 49.

195 SEPA 2017, p. 9.

196 Regeringskansliet 2013, p. 3.

197 Kischel 2015, p. 626.

198 Ibid, p. 617f.

199 Kischel 2015, p. 618f.

200 Ibid, p. 626.

instruments, depending on the enacting body.[201] Notably, Sweden does not possess a legal code in the sense of the BGB, a feature often highlighted in discussions of the Scandinavian legal system. Instead, Sweden has law codes (*balkar*), which typically do not focus on a single legal matter. These codes amalgamate a set of relevant laws pertaining to a particular subject into a comprehensive work.[202] this characteristic is not of substantial practical significance for legal comparativists; it more so illustrates the diverse interpretations of the term "code".[203] Furthermore, the statutory regulation of civil law in Sweden is characterised by its fragmented nature.[204]

An essential aspect of this research is the focus on Swedish tenancy law. The two primary sources of tenancy law are the Tenancy Act (Hyreslagen) incorporated in Chapter 12 of the Land Code (Jordbalken (1970:994)[205] and the Tenancy Bargaining Act (Hyresförnhandlingslag (1978:304)).[206] Given that cooperative rentals constitute a significant proportion of all rentals in Sweden (approximately 42% of dwellings in multi-dwelling buildings are owned by housing cooperatives),[207] the Cooperative Housing Act (Lag (2002:93) om kooperativ hyresrätt)[208] and the Cooperative Societies' Act (Lag (2018:672) om ekonomiska föreningar)[209] are also of considerable importance.

5.3.2.2 Comparative Research Questions

Is Any Level of Living Conditions, Including Thermal Efficiency, Guaranteed by the Legislator in Sweden?

Sweden, as a welfare state, boasts a well-developed social system that guarantees a basic level of security for its population, with individual rights playing a significant role in its

[201] Vogel in: Bogdan 2010, p. 20.

[202] Ring/ Olsen-Ring 2014, §2, Rn.36; Dotevall in: Smits 2006, p. 700; Carlson 2012, p. 36.

[203] Kischel 2015, p. 599f.

[204] Dotevall in: Smits 2006, p. 700.

[205] *Jordabalk(1970:994)*; Terner Center for Housing Innovation 2017, p. 9; BBSR 2016b, p. 61.

[206] Hyresförhandlingslag. n.d. Accessed 24 December 2023. https://www.riksdagen.se/sv/dokument-och-lagar/dokument/svensk-forfattningssamling/hyresforhandlingslag-1978304_sfs-1978-304/.; Swiss Institute of Comparative Law 2016, p. 58.

[207] 'Just over 5 Million Dwellings in Sweden'. n.d. Statistiska Centralbyrån. Accessed 24 December 2023.https://www.scb.se/en/finding-statistics/statistics-by-subject-area/housing-construction-and-building/housing-construction-and-conversion/dwelling-stock/pong/statistical-news/dwelling-stock-2020-12-31/.

[208] Lag om kooperativ hyresrätt. n.d. Accessed 24 December 2023. https://www.riksdagen.se/sv/dokument-och-lagar/dokument/svensk-forfattningssamling/lag-200293-om-kooperativ-hyresratt_sfs-2002-93/.

[209] Lag om ekonomiska föreningar. n.d. Accessed 24 December 2023.https://www.riksdagen.se/sv/dokument-och-lagar/dokument/svensk-forfattningssamling/lag-2018672-om-ekonomiska-foreningar_sfs-2018-672/.

formation.[210] This system is based on the principle of universality, meaning that these benefits are viewed as social rights available to the entire population, not just economically disadvantaged or vulnerable groups. This universality distinguishes the Swedish welfare system from selective welfare policies that cover only individuals struggling to meet their basic needs independently.[211] The focus of the Swedish social system includes social protection and welfare systems, employment, working conditions, wage-setting systems, and industrial relations.[212] The social welfare legislation is well developed and includes three major groups of expenses to be covered:

1) Health and medical services;
2) Social services for individuals in need of temporary or permanent support or assistance and
3) Services and insurances, including benefits from social insurances, social benefits, and financial allowances.[213]

Swedish housing policy also contributed to the development of the Swedish welfare model.[214] While there is no specific right to housing under Swedish national law,[215] the concept of home, as understood by Swedish courts, is often interlinked with property interests. This perspective is influenced by references to *Hans Danelius*, a former Justice of the Supreme Court of Sweden, who wrote that the right to respect one's home is more limited if the individual does not own it but merely rents or leases.[216]

However, the Instrument of Government,[217] one of Sweden's four constitutional documents, stipulates in Chap. 1, Article 2 that "Public power shall be exercised with respect for the equal worth of all and the liberty and dignity of the individual. The individual's personal, economic, and cultural welfare shall be fundamental aims of public activity. In particular, the public institutions shall secure the right to employment, housing and education, and promote social care and social security, as well as favourable conditions for good health."[218] Although this article mentions the right to housing, its wording suggests

---

[210] Anxo 2015, p. 9.

[211] Rothstein 2010, pp. 30–32; Swedish Ministry of Finance. 2017. 'The Swedish Model'. 20 June 2017. https://www.government.se/information-material/2017/06/the-swedish-model/., p. 15; Behrendt 2002, p. 17.

[212] Anxo 2015, p. 9.

[213] Bogdan 2010, p. 119.

[214] Bååth 2014, p. 34.

[215] Grundberg 2016, p. 50.

[216] Danelius 2015, p. 427; Baheru 2017, p. 164.

[217] Svensk Författningssamling. n.d. http://rkrattsdb.gov.se/SFSdoc/03/030593.PDF.

[218] The Instrument of Government. n.d. Accessed 24 December 2023. https://www.riksdagen.se/globalassets/07.-dokument--lagar/the-instrument-of-government-2015.pdf.

that it is more an objective goal for society than an enforceable right.[219] A historical approach to legal interpretation, particularly the examination of the preparatory work for this provision, reveals that Article 2 was included to align with international conventions ratified by Sweden, including the UN Declaration of Human Rights. This paragraph declares specific values as fundamental, yet legally binding basic rules cannot protect them for various reasons.[220]

Constitutional law does not hold as central a role for Swedish legal scholars as it does in Germany, where the Federal Constitution (Art. 1 (1) in conjunction with Art. 20 (1) GG) guarantees a minimum standard of living, as previously analysed. German legal scholars might be inclined to view other legal systems through the lens of their national legal order, leading them to seek a similar approach in Swedish legislation. However, this proves challenging due to the unique interpretation of fundamental rights within the Swedish legal order.[221]

In Sweden, fundamental rights are perceived as inalienable and universal, existing independently of the standard legal order. For instance, human dignity is not viewed as a subjective fundamental right, but rather as a crucial foundation for the construction of a democratic society. The most effective protection of human dignity, and other rights derived from it, is considered to be the democratic system itself. According to this perspective, even the most robust constitutional protection of rights and freedoms is insufficient if the democratic consciousness of citizens is compromised. Thus, Article 2 of the Instrument of Government does not offer absolute constitutional protection but is seen as exerting a protective influence through its psychological impact on society. Nonetheless, the concept of human dignity is a cornerstone of several Swedish legal frameworks, including the Social Services Act, the Health Care Act, and various medical declarations and codes.[222]

The proactive debate on the right to excellent or adequate housing began in Sweden around the mid-2010 s. In 2015, approximately 5,000 vulnerable EU citizens were either homeless or living in informal settlements within the country. By 2017, the number of homeless individuals had doubled compared to 1993.[223] In contrast to many neighbouring countries, in Sweden, it is primarily the responsibility of municipalities to provide housing for their inhabitants. During this period, municipalities lacked a uniform approach to addressing this issue: while some opened shelters and campsites for these individuals, others merely offered tickets back to their countries of origin. This situation

219 Grundberg 2016, p. 50.

220 Statens medicinsk-etiska råd 2012, p. 123.

221 Kischel 2015, p. 617.

222 Statens medicinsk-etiska råd 2012, p. 123f..

223 Knutagård/ Heap/ Nelson 2019, p. 4.

led to a political debate over whether vulnerable EU citizens have a right to housing in Sweden,[224] thereby intensifying the discussion about adequate housing.

As of the time of writing, Sweden had not implemented a national homelessness strategy. One contributing factor to the rising homelessness and housing exclusion is the country's scarcity of affordable housing. Additionally, the high costs associated with producing new housing exacerbate this issue. This problem is linked to another challenge: the rents in newly constructed housing are determined outside the traditional user value system of the Swedish rental market, leading to a significant increase in rental costs. The Swedish housing market is highly deregulated, and unlike in many countries, it does not operate on profit-based and competitive market principles.[225] Furthermore, Sweden does not have a social housing sector specifically reserved for vulnerable social groups. Instead, dwellings are available for rent to anyone, regardless of income. A substantial portion of rental housing in Sweden is provided and managed by non-profit municipal housing companies.[226]

As discussed in Chap. 3, the prevalence of energy poverty in Sweden is minimal, thanks to the country's robust social policies. These policies effectively address energy poverty, maintaining it at low levels. A strong social support system, comprising various social benefits, assists affected individuals in covering their essential living expenses, including housing, energy, and electricity costs.[227]

To qualify as a vulnerable person or household and receive social benefits, specific criteria must be met. Firstly, the potential recipient's income, including financial assets, must be below a certain minimum threshold set by the government. Secondly, entitlement to certain benefits, such as unemployment benefits, sickness pay, and pensions, requires the individual to have previously paid premiums. Thirdly, membership in specific social groups is a prerequisite for entitlement to benefits like child benefits, pensions, and benefits for people with disabilities.[228] Moreover, non-citizens must reside in Sweden for more than one year to be eligible for any social transfer, as stipulated in Chap. 4, Sections 1 and 3 of the Swedish Social Services Act.[229]

Additionally, the Swedish legislation imposes further conditions for individuals to be entitled to specific social benefits. For instance, a potential recipient of a subsistence allowance under Chap. 4, Section 1 of the Swedish Social Services Act must be available for the labour market if capable of working. This availability may require participation in municipal adult education programs in Swedish for immigrants or equivalent education

[224] Grundberg 2016, p. 5.

[225] Knutagård/ Heap/ Nelson 2019, p. 4.

[226] Chen/ Öst 2005, p. 606f.

[227] EPOV 2020, p. 109.

[228] Janson 2002, p. 2.

[229] Socialtjänstlag (2001:453). n.d. Accessed 24 December 2023. https://www.riksdagen.se/sv/dokument-och-lagar/dokument/svensk-forfattningssamling/socialtjanstlag-2001453_sfs-2001-453/.

at a folk high school. Only under justifiable circumstances can an individual be entitled to a subsistence allowance even if they are not available for the labour market. The assistance provided is designed to ensure a reasonable standard of living and to enhance the individual's ability to live independently. This includes covering reasonable costs for food, clothing, recreational activities, consumables, health and hygiene products, newspapers and phone, as well as housing, household electricity, travel to work, home insurance, and membership in a labour union or unemployment fund, as outlined in Chap. 4, Section 3 of the Swedish Social Services Act.[230] This social benefit seems to be similar to the former German ALG II.[231]

Further, the Swedish welfare system includes a housing allowance as a specific social benefit. This allowance offers financial support to households in need, targeting not only low-income groups but also other vulnerable categories such as chronically ill or disabled persons and households with children. It covers housing costs, which include heating expenses based on the temperature zone and the floor area of the residence.[232] In 2019, a total of 149,306 households in Sweden benefited from this type of social support.[233]

The distribution of housing allowance is based on the principle of "demonstrable and abject need". To qualify, the recipient's income, including financial assets, must be below a threshold established by governmental bodies.[234] This income level is calculated as the pre-tax income, excluding any social benefits.[235] The housing allowance is intended to cover reasonable expenses for housing and heating, as stipulated in Chap. 4 , Section 3 of the Social Services Act. Additionally, Sweden offers a social assistance measure that helps cover the living expenses of low-income individuals, including accommodation and electricity costs.[236]

Unlike many other countries, the Swedish housing allowance system is accessible to all tenure types, without favouring any particular form of tenure. In the Swedish housing market, there are three major tenure forms: owner occupancy, tenant-ownership,[237] and

---

[230] Own translation.

[231] Now *Bürgergeld* (citizens' income).

[232] EPOV 2020, p. 109.

[233] 'Sweden: Housing Allowance Recipients by Type 2021'. n.d. Statista. Accessed 24 December 2023. https://www.statista.com/statistics/536899/sweden-housing-allowance-recipients-by-type-of-household/.

[234] Janson 2002, p. 2.

[235] Chen/ Öst2005, p. 608.

[236] EPOV 2020, p. 109.

[237] In Sweden, tenant-owned housing, also known as cooperative housing, refers to multi-family apartments with a distinctive feature: the members do not obtain full ownership rights of the dwellings they inhabit. However, they have the flexibility to sell their tenant-owned dwellings at any time, with no limitations on the selling price in most cases. There are a few exceptions where the

rental. As of 2017, the owner-occupied segment represented just over 40% of the housing stock.[238] The main task of the housing allowance system in Sweden is to support vulnerable social groups, particularly those with children, in achieving better living conditions. This includes access to appropriately sized dwellings at reasonable costs. Additional benefits of this system include reducing overcrowding and serving as a cost equaliser for housing across different cities and regions within the country.[239]

In summary, Swedish legislation does not guarantee a specific standard of living conditions, including thermal energy efficiency. However, the country's social legislation safeguards vulnerable social groups, enabling them to apply for social benefits. These include a housing allowance that covers their housing and energy costs. Nevertheless, there is no enforceable right for individuals to claim a specific level of energy efficiency in housing.

Before assessing the minimum guaranteed level of life conditions in Sweden, it is pertinent to provide a brief overview of Swedish tenancy law. As previously mentioned, municipalities bear the responsibility for housing provision in Sweden. Public housing, a key policy instrument to increase the supply of affordable housing, is also a cornerstone of the welfare state.[240] This system was developed under the governance of the Swedish Social Democratic Workers' Party (SAP) from the 1930 s until 2006. The Swedish housing policy, known as "*Allmännytta*", translates to "public utility" or "for the benefit of everybody" in English, reflecting the government's perspective on housing policy as a social right rather than a private concern.[241]

The Swedish housing regime, in general, is often described as a "corporatist-housing regime",[242] with public housing being one of its main pillars. Municipal housing companies hold a substantial portion of the rental housing market. Approximately one-fifth of the Swedish housing stock is public, managed by local Municipal Housing Companies (MHCs). Public housing in Sweden is accessible to all, without distinguishing between social housing and other types. There is a direct social housing stock, created as part of additional supply-side intervention, but its proportion is relatively small, amounting to about 1% of the total housing stock. These properties are leased by social authorities, who then sublet them to households facing "social problems".[243] Given the lack of a

housing associations have the authority to intervene and set the selling prices. As of 2017, cooperative housing constituted approximately 23% of Sweden's national housing stock, see. Terner Center for Housing Innovation, p. 7f., Chen/Öst 2005, p. 606f.

[238] Terner Center for Housing Innovation 2017, p. 8.

[239] Chen/ Öst 2005, p. 607.

[240] Hananel/ Krefetz/ Vatury 2018, p. 462.

[241] Hananel/ Krefetz/ Vatury 2018, pp. 464f.

[242] Knutgård/Kristiansen, p. 97; Bengtsson 2006; Hananel/ Krefetz/ Vatury 2018, pp. 462ff.

[243] Terner Center for Housing Innovation 2017, p. 6.

clear distinction between social housing and other housing types, the general provisions of Swedish tenancy law will be explored to address the research question.

In Sweden, there is a clear distinction between private and municipal tenancy. The rent control laws are notably tenant-friendly. Annually, negotiations usually occur between landlords and tenant representatives—tenant associations for both private and public housing stock at the local level, predominantly involving the Swedish Union of Tenants. These negotiations revolve around two main aspects: the utility value of dwellings and changes related to the cost of building management.[244] The Swedish Union of Tenants is one of the strongest (if not the strongest) tenant organisations in the world.[245]

In 2002, a new legal institute—the so-called cooperative rentals (*kooperativ hyresrätt*)—was introduced to encourage the construction of new rental properties and enhance tenants' influence. Cooperative rentals represent a hybrid model, blending elements of housing cooperatives and classical residential rents. While these models share many similarities, they also have key differences described further.[246] The landlord in this model is a cooperative rental association, with the tenant being a cooperative member. The rental association may either own the residential unit or lease it, typically from the municipality. The association members determine the rent of the cooperative dwelling through bylaws, and these rent levels are usually established outside the usual negotiation procedure regulated for residential rents, as set out in the Rent Negotiation Act (*Hyresförhandigslag 1978:304*)[247] and the Land Code (Chapter twelve) (*Jordbalken* 1970:994).[248] Most provisions of the tenancy law (the Land Code), except those concerning remedial injunctions, tenants' influence, and utility value, apply to cooperative tenants. However, individual cooperative tenants are barred from approaching the Regional Rent Tribunal to assess the fairness of their rent.[249] Cooperative rentals are also regulated by the Cooperative Housing Act (*Lag 2002:93 om kooperativ hyresrätt*)[250] and the Cooperative Societies' Act (*Lag 2018:672 om ekonomiska föreningar*).[251] Some relevant

---

[244] Terner Center for Housing Innovation 2017, p. 9; BBSR 2016b, p. 61.

[245] Rolf 2021, p. 186.

[246] Carlson 2012, p. 437; Lunderup 2016, p. 1.

[247] Hyresförhandlingslag. n.d. Accessed 24 December 2023. https://www.riksdagen.se/sv/dokument-och-lagar/dokument/svensk-forfattningssamling/hyresforhandlingslag-1978304_sfs-1978-304/.; Swiss Institute of Comparative Law 2016, p. 58.

[248] *Jordabalk(1970:994*; Terner Center for Housing Innovation 2017, p. 9; BBSR 2016b, p. 61.

[249] Carlson 2012, p. 438.

[250] Lag om kooperativ hyresrätt. n.d. Accessed 24 December 2023. https://www.riksdagen.se/sv/dokument-och-lagar/dokument/svensk-forfattningssamling/lag-200293-om-kooperativ-hyresratt_sfs-2002-93/.

[251] Lag om ekonomiska föreningar. n.d. Accessed 24 December 2023. https://www.riksdagen.se/sv/dokument-och-lagar/dokument/svensk-forfattningssamling/lag-2018672-om-ekonomiska-foreningar_sfs-2018-672/.

provisions are incorporated into the Housing Management Act (*Bostadsförvaltnigslagen 1977:792*)[252].

At first glance, the utility value system in Swedish tenancy law appears similar to the German *Mietspiegel*, meaning that rents are determined by comparing them with similar tenancies in the neighbourhood.[253] The Swedish rental market features a comprehensive rent control system built on two main pillars: a significant share of dwellings provided by public housing and a well-established tenants' association representing tenants in rent negotiations. This system helps keep rents below market levels in Sweden.[254]

Regional Rent Tribunals play a crucial role in this system. Their responsibilities include managing the utility value system and setting limits for "reasonable rents", which are based on the highest rate of the rental price for similar apartments plus an additional 5%. These tribunals also oversee negotiations between landlords and the Swedish Union of Tenants.[255] When a landlord wishes to alter rental terms, they must send a written notice to the union, thereby initiating the negotiation process. Tenants are generally bound by the newly negotiated terms of the rental contract if it contains a bargaining clause. However, they have the option to approach the rent tribunal to request amendments to the contract.[256]

In cases where no agreement is reached, the Rental Market Committee intervenes to find a resolution.[257] If the rental contract lacks a bargaining clause, any rent increase must be negotiated individually. Should the parties fail to reach an agreement on the rent increase, the rent tribunal must review the case. Additionally, there is a six-month prohibition against rent increases following the last increment.

For a tenancy agreement concluded according to the negotiation procedure to be binding on the tenant, it must include a negotiation clause. Under §3 of the *Hyresförhandlingslag*, tenants who wish to negotiate directly with the landlord about tenancy terms can revoke the negotiation clause, provided the tenancy has lasted at least three months. Additionally, according to §2 of the Hyresförhandlingslag, the removal of a negotiation clause from the tenancy agreement is permissible in cases of doubt. According to §5 of the *Hyresförhandlingslag*, once a negotiation agreement is signed, the landlord's primary obligation is to initiate negotiations with the tenants' association independently on the following matters:

---

252 *Hyresförhandlingslag. n.d. Accessed 24 December 2023.* https://www.riksdagen.se/sv/dokument-och-lagar/dokument/svensk-forfattningssamling/hyresforhandlingslag-1978304_sfs-1978-304/.

253 Bååth 2014, p. 5.

254 Farahani 2017, p. 17.

255 Terner Center for Housing Innovation 2017, p. 9; Bååth 2014, p. 5.

256 Bååth 2014, p. 5.

257 Terner Center for Housing Innovation 2017, p. 9.

- Rent increase;
- Setting the amount of rent for a new tenant if the demanded rent exceeds the previous rent;
- Changing the basis for calculating charges for heating, water, electricity, or water supply as per *Jordbalken* 12:19; and
- Modifying the tenant's right to use common areas.[258]

These negotiation processes offer individual tenants security in the event of a rent increase. Another mechanism of rental protection in Sweden is the "security of tenure". This principle entitles tenants to rent their dwellings as long as they are not in rental arrears and do not disturb their neighbours.[259] The termination of rental contracts is severely restricted, and landlords generally cannot terminate rental agreements unilaterally.[260] Although Swedish legislation does not require a specific form for rental agreements concerning dwellings, they must be in writing if requested by either the landlord or the tenant.[261]

The next question addresses the costs covered by social welfare benefits, i.e., the level of adequate housing provided for vulnerable Swedish tenants. This thesis will delve into social assistance in detail, as it is a social benefit not classified as part of social insurance in Sweden.[262] It provides only the minimum security level for cases where other means of economic support are inapplicable. Social assistance is distinct from other social security benefits and does not depend on the individual contributions of the applicant.[263] Before applying for this type of social assistance, the individual must claim all other government-financed allowances[264] and exhaust their savings to become eligible.[265]

Sweden offers a housing allowance as part of its family allowance scheme, regulated by Chapters 93 to 103e of the Swedish Social Insurance Code.[266] According to Chapter 95 Section 2, the allowance encompasses various types, including: an allowance for families with children living in the same household; an allowance for individuals aged between 18 and 29 without children; an allowance for families where the children predominantly

258 Hyreskomissionen 2018, p. 20.

259 Terner Center for Housing Innovation 2017, p. 9; BBSR 2016b, p. 61.

260 BBSR 2016b, p. 61.

261 Carlson 2012, p 417f.

262 European Commission 2013, p. 6.

263 Behrendt 2002, p. 177.

264 OECD 2020, p. 10.

265 ESPN 2016, p. 8.

266 Socialförsäkringsbalk (2010:110). n.d. Accessed 24 December 2023. https://www.riksdagen.se/sv/dokument-och-lagar/dokument/svensk-forfattningssamling/socialforsakringsbalk-2010110_sfs-2010-110/.

reside in one household; and an allowance for parents living part-time with their child.[267] The housing allowance is granted as a subsidy to cover housing costs, with the amount depending on the individual's income and other circumstances detailed in the Social Insurance Code. It is issued on a temporary basis, calculated from an estimated basic income, and determined retrospectively based on the assessed income upon which the grant is based (Chapter 95 Section 4, Social Insurance Code).[268] This allowance is similar to the German *Wohngeld* and aims to support specific social groups in achieving better living conditions. Therefore, it will not be examined further in this analysis.

Social assistance, known in Sweden as *Ekonomiskt bistånd* or *försörjningsstöd,* includes both a national standard (*riksnorm*) and the coverage of reasonable expenses, such as housing costs. The national standard is a fixed sum, recalculated annually by the government based on family budget studies and price inflation. This sum serves as a guideline, allowing municipalities the discretion to provide a higher (but not lower) level of support depending on the individual circumstances of the applicant.[269] The coverage of reasonable costs encompasses not only rental expenses but also electricity bills and home insurance, with the specific amounts varying by municipality.[270]

To analyse which housing costs can be deemed reasonable, it is necessary to ascertain the lowest acceptable standard of living. Swedish tenancy law stipulates the minimum acceptable standard of living in Section 18a of the Tenancy Act. According to this provision, landlords are obligated to ensure that dwellings are in a condition that renders them fully usable for their intended purpose. This standard is determined based on the general expectations prevalent in the area where the dwellings are located, taking into account the date when the tenant takes possession of the property.[271] The minimum acceptable standard of living, as defined by Swedish law, includes the provision of necessary equipment within the dwelling. This encompasses facilities for continuous heating, a consistent supply of hot and cold water, electricity for typical household consumption, wastewater drainage, and a personal hygiene unit comprising a toilet, a washbasin, and a bathtub or shower. Additionally, the dwelling must include a cooker, sink, refrigerator, adequate storage space, and worktops.[272]

If a dwelling fails to meet the lowest acceptable standard of living due to the intended rental purpose, tenants have the right to approach the Regional Rent Tribunal. The tribunal may issue an order for improvements, enforceable through a penalty or fine, compelling the landlord to upgrade the dwelling to the requisite standard.[273] Should the landlord

---

[267] European Commission 2013, p. 29.

[268] Own translation.

[269] Hansen/ Schultz-Nielsen 2015, p. 6; OECD 2020, p. 10; ESPN 2016, p. 7.

[270] Hansen/ Schultz-Nielsen 2015, p. 6.

[271] Carlson 2012, p. 419.

[272] Carlson 2012, p. 421f.; Bååth 2014, p. 59.

[273] Carlson 2012, p. 421.

fail to comply with this order, the rent tribunal is empowered to appoint a trustee for compulsory management of the property, typically for a period of three to five years.[274] During this time, the trustee is responsible for collecting rent and conducting the necessary repairs. If the rent collected is insufficient to cover the costs of these repairs and there is no feasible option to secure a loan for this purpose, the building may be subject to demolition. In such cases, tenants are entitled to seek compensation for their damages from the landlord.[275]

Regarding the reasonable housing costs covered by social assistance, these vary depending on the municipality. However, the Swedish legislator has established the lowest acceptable standard of living within the Tenancy Act. Consequently, the housing costs for dwellings that meet this standard are presumed to be reasonable under Swedish law. Therefore, there is at least a basic level of security in Sweden related to the provision of adequate housing.

Are there any Legal Barriers Related to the Decarbonisation of Rented Dwellings in Sweden?

Sweden has effectively implemented various policies and measures to reduce both direct and indirect greenhouse gas emissions, particularly within its building stock.[276] One of the key mechanisms to promote energy efficiency is the implementation of energy and carbon dioxide taxation, which exceeds the minimum level stipulated in the Energy Taxation Directive. This approach incentivises the adoption of energy-efficient practices.[277] The Swedish Buildings Regulations[278] specify limits for specific energy use (measured in kWh/m^2 per year), average thermal transmittance (W/m^2K), and the average leakage of the building envelope ($l/s\ m^2$) for both new constructions and buildings undergoing modernisation.[279]

In Sweden, the predominant residential lease type is the all-inclusive rent model. In this system, landlords charge a lump sum that includes the base rent along with heating and/or hot water costs. Tenants make their rental payments directly to the landlord, who is then responsible for settling bills with utility companies. In multi-residential buildings, utility costs are apportioned based on the floor area occupied by each apartment unit. One of the disadvantages of this method is its lack of differentiation between corner

[274] Bostadsförvaltningslag (1977:792). n.d. Accessed 24 December 2023. https://www.riksdagen.se/sv/dokument-och-lagar/dokument/svensk-forfattningssamling/bostadsforvaltningslag-1977792_sfs-1977-792/.; Bååth 2014, , p. 60.

[275] Bååth 2014, p. 60.

[276] Sweden Draft NECP, 2021, p. 15.

[277] Council Directive 2003/96/EC of 27 October 2003 restructuring the Community framework for the taxation of energy products and electricity, OJ L 283, 31.10.2003, pp. 51–70.

[278] Om Boverkets Byggregler, BBR - Boverket. n.d. Accessed 24 December 2023. https://www.boverket.se/sv/byggande/regler-for-byggande/om-boverkets--byggregler-bbr/#.

[279] Sweden Draft NECP 2021, p. 15.

apartments, which may have higher heating requirements, and apartments surrounded by other heated spaces.[280] Despite this, Swedish landlords are incentivised to enhance their buildings' energy efficiency to reduce overall consumption. However, as tenants pay a fixed cost for energy, there is a lack of direct incentives for them to conserve energy. Landlords may set consumption limits, such as regulating the maximum indoor temperature during winter, to encourage more efficient energy use.[281]

Energy performance have been a long-standing component of Swedish building regulations.[282] Since the 1960 s, Sweden has implemented building regulations that set minimum requirements for energy use in new buildings.[283] However, these codes and standards often lag behind technological advancements, posing a barrier to the adoption of energy-efficient innovations.[284] Researchers and constructors have critiqued these regulations for having limited impact on achieving Sweden's energy goals, citing the relatively low level of energy requirements.[285]Furthermore, these legal norms predominantly affect new buildings, leading to a challenge in decarbonising the building sector, particularly residential buildings. This is because the existing building stock is only slowly being replaced by new, more energy-efficient buildings, and as building technologies continue to improve, the energy performance of existing buildings remains crucial.[286] Further, there is a noted energy-efficiency gap in Sweden between optimal and actual energy consumption levels. This gap is attributed to structural and market barriers that impede achieving optimal energy usage.[287]

The analysis of the Swedish Tenancy law reveals that energy modernisation is not explicitly regulated within its scope. This was further confirmed in an interview conducted by the author of this thesis in January 2022 with *Boverket*—the Swedish National Board of Housing, Building and Planning. Notably, the term "energy modernisation" is not commonly used in Swedish legislation. Moreover, Swedish legislation is not familiar with the term "energy modernisation" at all.[288] Nonetheless, existing rules pertaining to housing renovation are applicable to energy modernisation. In cases of housing renovation, tenants are entitled to a rent reduction for the duration of the renovation. If the dwellings become uninhabitable during renovations, landlords are obliged to provide tenants with an alternative suitable apartment. In situations where tenants refuse to

[280] JRC 2014, p. 23f.

[281] Ibid.

[282] IEA Sweden 2019, p. 15.

[283] Naturvardsverket 2021, p. 25.

[284] Persson/ Grönkvist 2015, p. 298.

[285] Persson/ Grönkvist 2015, pp. 299f.

[286] IEA Sweden 2019, p. 15.

[287] Persson/ Grönkvist 2015, p. 296.

[288] During the interviews with the Swedish stakeholders, all three respondents asked to specify what is meant under the term "energy modernisation".

consent to renovation measures, which the landlord must announce in advance in writing, and no agreement is reached, the landlord can file a claim in the rental court. This step can be taken at the earliest two months after the tenant receives the written notification. The court may then override the need for the tenant's consent if the refusal is deemed unjustified.[289]

Under Swedish tenancy law, landlords are entitled to carry out improvements that increase the utility value of a dwelling, including energy modernisation, as stated in Section 18d of the Tenancy Act. However, these improvements can only proceed with the approval of the concerned tenants. If tenants withhold their approval, the landlord may seek permission from the rent tribunal, especially in cases where the proposed measures are of considerable interest and do not impose undue burden on the tenants. In this process, both the individual interests and circumstances of the tenants and the landlord's interest in executing modernisation measures are taken into account. The aim is to strike a balance between these potentially conflicting interests. Importantly, if a modernisation measure elevates the dwelling's standard to the minimum acceptable level as specified in Section 18a of the Tenancy Act, the landlord's interest takes precedence over the tenants' objections, meaning tenants cannot oppose the renovation.[290]

Landlords are required to inform tenants in writing about any planned measures, as per Section 18e of the Tenancy Act. If an agreement on the measures cannot be reached, the landlord must wait two months before applying to the rent tribunal. Tenants are entitled to claim compensation from the landlord for disturbances caused by renovation measures unless this right has been waived in their tenancy agreement.[291]

Additionally, landlords are responsible for all maintenance works and repairs necessary to keep the dwelling in a fully serviceable condition for its intended purpose, as outlined in Sections 15 and 9 of the Tenancy Act. This includes carrying out customary repairs, such as wallpapering or painting, within a reasonable timeframe, particularly if the dwelling requires such maintenance.[292]

In Swedish tenancy law, a residential tenant is permitted to undertake minor modernisations of the dwelling at their own expense, such as wall-painting, wallpapering, and similar activities. However, if these alterations result in a decrease in the utility value of the dwelling, the tenant is liable to compensate the landlord for any resultant damages. Judicial review regarding any decrease in utility value is typically conducted only after the tenant vacates the property.[293] If the dwelling is not in the condition stipulated in the tenancy contract, the tenant is entitled to report these defects. The landlord is then obligated to rectify these issues within a reasonable period, depending on the nature

[289] BBSR 2016b, p. 91.

[290] Bååth 2014, pp. 91f.

[291] Ibid. p. 92.

[292] Ibid, p. 83.

[293] Ibid, p. 55.

of the defect. Should the landlord fail to address these concerns, the tenant may carry out the necessary repairs themselves, as per Section 11 paragraph 1 of the Tenancy Act. However, this "right to self-care" is seldom exercised in Sweden, particularly if the defect does not require immediate intervention, due to the associated risks for the tenant.[294]

Swedish Tenancy Law appears to be relatively favourable to landlords, especially those undertaking energy modernisation to meet minimum energy efficiency standards. In such cases, the landlord's interests in carrying out the modernisation typically outweigh any opposing interests of the tenants, should they not approve of these measures. Nonetheless, a major impediment to energy modernisation in Sweden is the rent control system, which has been heavily criticised.[295] Rent control withdraws the landlord's incentive to maintain and improve the dwellings within the tenancy.[296] Furthermore, the prevalent all-inclusive rent model contributes to "split incentives" for tenants, as they do not directly benefit from reduced heating costs following energy modernisation.[297] This lack of direct incentives can lead to rebound effects.[298]

Overcoming the Barriers Related to Decarbonisation of the Rented Dwellings in the Swedish Legal Order

Swedish tenancy law is notably tenant-friendly,[299] with tenancy agreements typically being open-ended, unless a specific time limit is agreed upon.[300] Landlords are only entitled to terminate rental agreements under very limited conditions, explicitly listed in *Jordbalken* (1970:994) Sections 12:42 and 12:46. These conditions generally stem from a gross breach of duty by the tenant.

As previously mentioned, distinctive features of the Swedish housing market include collective rent agreements and all-inclusive rents. Tenants pay a gross rent which encompasses the net rent minus any rent losses, operating costs, maintenance costs, and financial expenses. Operating costs cover maintenance, repairs, insurance, any ground rent, and property taxes. Heating, electricity, gas, water, sewage, and waste disposal are included in these costs, though they are tariff-related. Financial expenses encompass interest payments and any applicable guarantee fees. All these expenses are included in the basic rent (also known as "warm rent"), except for individually metered and billed electricity and water.[301]

---

[294] Ibid, p. 80.

[295] Norberg/ Juul-Sandberg 2016, p. 3; Andersson/ Södeberg 2012, p. 162.

[296] Arnott 2003, p 99.

[297] Agora Energiewende und Universität Kassel 2021, p. 6.

[298] Ibid, p. 7.

[299] Terner Center for Housing Innovation 2017, p. 9; BBSR 2016, p. 61.

[300] Baheru 2017, p. 14.

[301] Hyreskommissionnen 2018, p. 18.

The rent control system in Sweden ensures that rents remain below market prices. Consequently, the allocation of rental housing is not solely based on a tenant's willingness to pay for a particular flat but is also influenced by their position in the housing agency queue. The longer a person has waited, the greater the selection of flats available to them.[302] This system facilitates access to better dwellings for vulnerable tenants.

Dating back to 1942, the Swedish rent control system was initially established to prevent rent increases during times of crisis.[303] It was amended in the 1950 s with the introduction of the Security of Tenancy Act (*Besittningskyddslagen 1956:568*), shifting from a fixed rent price set by authorities to a rent ceiling based on price comparisons with equivalent dwellings. In the 1970 s, this system evolved into the current "use value system",[304] where rents are set based on the utility value of the dwelling and are cost-based, meaning they cannot exceed rents for equivalent dwellings. A unique aspect of Swedish tenancy law is the collective bargaining for basic rents by the Swedish Tenants' Association, representing tenants and landlords, including municipal housing companies.[305] Thus, the basic rent price in municipal housing depends on the outcome of these negotiations and also influences rents set by private landlords. Tenants in Sweden benefit from a degree of stability when signing a rent agreement, as they are shielded from unexpected rent increases. They also have the advantage of not having to worry about energy costs, which is particularly beneficial for vulnerable groups during colder months. Additionally, tenants can challenge any rent increase at the tenancy court, providing further protection.[306]

As previously mentioned, there is no formal legal definition for 'energy modernisation,' nor is there a provision that allows landlords to increase rent specifically due to energy modernisation of a dwelling under Swedish law. This scenario presents a landlord-tenant dilemma: while tenants are better protected against rent increases, landlords have greater incentives to undertake energy modernisation. However, significant barriers to the decarbonisation of the rented dwelling stock include high investment costs for energy modernisation and a general lack of financial resources among landlords and homeowners, as indicated by various studies.[307]

Furthermore, the concept of all-inclusive rents in Sweden alleviates some of the tension inherent in the landlord-tenant relationship. Sweden is unique in Europe for generally basing contractual rents on gross warm rent, where landlords are responsible for the total cost of heating and hot water.[308] This approach aligns with Art. 9 (1) of the

[302] Wilhelmsson, 2021, p. 3

[303] Fritzon 2020, p. 11.

[304] Wilhelmsson 2021, p. 8

[305] Fritzon 2020, p. 11.

[306] Ibid.

[307] Pardalis et al. 2019, p. 4.

[308] BBSR 2016, p. 96; Agora Energiewende and Universität Kassel 2021, p. 6.

EED, which mandates member states to ensure that energy costs are reflected through competitively priced individual meters. These meters should accurately represent the final customer's actual energy consumption and provide information on the actual time of use, as long as this is technically feasible, financially reasonable, and proportionate to the potential energy savings. Art. 9 (2) sub-paragraph 2 of the EED allows member states to forego the installation of individual consumption meters if it is not technically feasible or cost-efficient, permitting cost-efficient alternative methods of heat consumption measurement in such cases. In Sweden, the adoption of individual measuring systems is often viewed as a socially burdensome cost factor.[309]

The all-inclusive rent system in Sweden incentivizes landlords to undertake energy modernisation as it directly reduces their operating costs for heating and hot water. This effectively addresses a significant aspect of the landlord-tenant dilemma, motivating landlords to refurbish dwellings to save on heating costs. Additionally, refurbishment measures aimed at reducing energy consumption are not grounds for rent increases, which creates a strong incentive for landlords to implement cost-effective renovation measures without imposing additional financial burdens on tenants. This approach is further bolstered by Sweden's relatively high CO_2 tax, which surpasses the fixed prices for emission trading set by the BEHG until 2025. Landlords are thus encouraged to reduce not only their heating costs but also their CO_2 tax liabilities through energy modernisation.[310] In the long-term perspective, it is conceivable that Sweden may transition from its all-inclusive rent system to a net cold rent model to align with the requirements of the EED.[311]

The fixed warm rent system offers several advantages in resolving the landlord-tenant dilemma. It elevates the economic potential in the rental market by aligning the cost-benefit calculations of landlords with those of owner-occupiers. Combining this system with an expanded emission trading system that includes the building sector could yield even better results. Sweden's positive experience with this approach indicates that landlords, benefiting from the reduced burden of CO_2 pricing and lower heating costs, are increasingly incentivised to undertake ambitious energy-saving measures.[312]

However, there is a notable disadvantage. Shifting the entire heating cost burden to a fixed warm rent creates a usage-related split incentive. Tenants no longer have a financial incentive to conserve heating energy, potentially leading to economically careless heating practices.[313] Moreover, tenants do not directly benefit from the cost savings associated with energy renovations in terms of allocational costs distribution.[314]

309 BBSR 2016, p. 61.

310 Agora Energiewende and Universität Kassel 2020, p. 11.

311 BBSR 2016, p. 96.

312 Agora Energiewende und Universität Kassel 2020, p. 16.

313 Klinski/ Keimeyer/ Braungardt 2021, p. 4.; JRC 2017, p. 6.

314 BBSR 2021, p. 85.

5.4 Comparing Similarities and Differences of Both Countries

Both Sweden and Germany, as welfare states, exhibit strong social legislation aimed at providing assistance to those in need. However, in both countries, energy poverty is not directly addressed in the legislation. It appears to be a latent issue, with the presumption in the social legislation being that energy poverty does not significantly affect these nations.

In terms of tenancy law, both countries have designed their systems to be tenant-friendly. However, the all-inclusive rent system and the rent price bargaining mechanism in Sweden are unique features that do not have equivalents in German tenancy law. Swedish tenancy law offers tenants, especially those in municipal housing, greater protection from unexpected rent increases. Additionally, landlords in Sweden are more incentivised to undertake energy modernisation as they are responsible for energy bills.

Unlike Germany, where the term "energy modernisation" is explicitly recognised and regulated in tenancy law, detailing the rights and obligations of both landlords and tenants during such modernisation processes, Swedish law does not specifically use this term. In Sweden, general norms for modernisation apply, covering the relevant aspects. However, in Germany, despite the legal recognition of energy modernisation, there is no explicit right to live in an energy-efficient dwelling.

A critical concern for vulnerable social groups, particularly those reliant on social assistance, is that the state typically only covers minimum rent costs. Consequently, low-income tenants in both countries often reside in the most affordable housing available, which tends not to be energy efficient. Nonetheless, due to Sweden's rent control system, low-income tenants may have better access to higher-quality housing compared to their counterparts in Germany.

5.5 Critical Evaluation of Replicability of the Warm Rent Model to Germany

The Swedish fixed warm rent model effectively addresses the landlord-tenant dilemma in Sweden. However, its replication in Germany encounters legal challenges, particularly with EU law and the Energy Efficiency Directive (EED). Art. 9b (3) EED obliges member states to ensure transparency and accuracy in billing for individual consumption. Therefore, billing must be based on actual individual consumption per Art. 10a (1) EED,[315] which is not feasible under a warm rent model.

In Germany, a genuine warm rent tenancy agreement is problematic due to national legal provisions. If such an agreement is in place, the Heating Costs Ordinance

[315] Becker Büttner Held 2020, p. 10.

(HeizkostenVO) is typically applicable, as per the prevailing legal opinion.[316] According to §4 of the HeizkostenVO, the building owner must measure the consumption of heat and hot water. The costs for heat and hot water supply are to be distributed among the users based on a mix of area and consumption under §§7 to 9a HeizkostenVO. Furthermore, §2 HeizkostenVO states that the ordinance's requirements prevail over the provisions of the tenancy agreement, except in cases specified in §11 HeizkostenVO.[317] These exceptions are beyond the scope of this thesis.

Therefore, heating costs cannot be included in the rent as a fixed sum without any calculation. The BGH ruled in 2006 that the provisions of the HeizkostenVO override any deviating contractual provisions without exception.[318] Although this ruling was initially in the context of a rent increase case, this legal opinion is predominant in the literature regarding future tenancy contracts.[319] The wording of the HeizkostenVO suggests that the legislator intended to allow warm rent contracts only in specific cases outlined in the ordinance (see §§2 and 11 HeizkostenVO). The legislator has explicitly chosen not to apply §§3 to 7 of the HeizkostenVO, which regulate the tracking and allocation of warm and hot water costs and handle related data.

According to the study of Agora, warm rents combined with steadily rising CO_2 prices on fuels are among four key prerequisites for a successful heat transition,[320] and thus, decarbonisation of the residential building sector. However, this study does not account for the non-legal context, particularly the significant difference in population size between Germany and Sweden. Germany's population is approximately eight times larger than Sweden's, which could lead to substantially higher energy usage by German tenants under a warm rent model, as they would not be directly responsible for heating costs. To make this model more environmentally friendly, a modified version of the gross warm rent model with direct feedback might be a viable solution, which will be explored in detail in the following chapter. In this model, landlords and tenants could agree on certain comfort conditions, such as a specified temperature level in the flat, covered by the landlord. Additionally, tenants could receive compensation for consuming less energy, introducing a direct feedback component.[321] However, this approach may

[316] Schumacher in: Theobald/Kühling 2021, HeizkostenVO, §2 Rn. 21; BayOLG, NJW-RR 1988, 1293 (1294); Flatow 2021, p. 866 ff.

[317] Flatow 2021, p. 866.

[318] BGH NZM 2006, 652; Flatow 2021, p. 867.

[319] Schumacher in: Theobald/Kühling 2021, HeizkostenVO; Flatow 2021, p. 866 ff; BeckOGK/Drager, Stand 1.4.2021, § 2 HeizkostenVO Rn. 3;

[320] Agora Energiewende. 2020. 'Warmmiete Macht CO2-Preis Auf Öl Und Gas Mieterfreundlich'. Pressemitteilung. https://static.agora-energiewende.de/fileadmin/Projekte/2020/2020_07_Warmmieten/2020-10-12_PM_Agora_Warmmieten.pdf.

[321] JRC 2017, p. 6.

not significantly benefit low-income tenants in Germany, whose energy costs are covered by the state.

These schemes, while redistributing the costs of energy modernisation to the landlord, do not necessarily incentivise landlords to decarbonise dwellings. The current cap on modernisation-related rent increases of 8% does not provide sufficient motivation for landlords to undertake such investments. Furthermore, information asymmetry persists, particularly in regions with less tight rental markets like Spree-Neiße, where landlords may have legitimate concerns about tenants moving out, leaving their properties vacant for extended periods. Additionally, the financial barriers for landlords, particularly those without available resources, are not addressed by this scheme. The cap on rent increases also renders renovation loans economically unfeasible due to extended repayment periods.

References

Abgeordnetenhaus Berlin. 2019. 'Vorlage – Zur Beschlussfassung – Gesetz Zur Neuregelung Gesetzlicher Vorschriften Zur Mietenbegrenzung, Drucksache 18/2347'. https://pardok.parlament-berlin.de/starweb/adis/citat/VT/18/DruckSachen/d18-2347.pdf.

Abramenko, Andrik. 2018. 'Mietpreisbremse – Ein Lehrstück Für Gesetz- Und Verordnungsgeber'. *Zeitschrift Für Rechtspolitik*, no. 2: 34–36.

Achenbach, Bruno. 2019. *Münchener Anwaltshandbuch Mietrecht*. Edited by Thomas Hannemann and Michael Wiegner. 5., Überarbeitete und erweiterte Auflage. München: C.H. Beck.

AG Mainz, 08.05.2013 – 86 C 323/11. 2013. Urt. AG Mainz.

Agora Energiewende. 2020. 'Warmmiete Macht CO_2-Preis Auf Öl Und Gas Mieterfreundlich'. Pressemitteilung. https://static.agora-energiewende.de/fileadmin/Projekte/2020/2020_07_Warmmieten/2020-10-12_PM_Agora_Warmmieten.pdf.

Andersson, Roland, and Bo Södeberg. 2012. 'Elimination of Rent Control in the Swedish Rental Housing Market: Why and How?' *Journal of Housing Research*, no. 2: 159–82.

Anxo, Dominique. 2015. 'Resilence and Success in Turbulent Times'. *Economia & Lavoro XLVIII*, no. 2: 9–29.

Arnott, Richard. 2003. 'Tenancy Rent Control'. *Swedish Economic Policy Review* 10 (January): 89–121.

Artz, Markus, and Ulf Börstinghaus. 2019. 'Artz/Börstinghaus: Das Am 1.1.2019 in Kraft Getretene Mietrechtsanpassungsgesetz'. *Neue Zeitschrift Für Miet- Und Wohnungsrecht*, no. 1: 12–25.

Börstinghaus, Ulf, and für. 2018. 'Entwurf Eines Gesetzes Zur Ergänzung Der Regelungen Über Die Zulässige Miethöhe Bei Mietbeginn Und Zur Anpassung Der Regelungen Über Die Modernisierung Der Mietsache (Mietrechtsanpassungsgesetz – MietAnpG)'. Neue Zeitschrift Für Miet- Und Wohnungsrecht, no. 12: 487–509.

Bååth, Olivia. 2014. 'Tenant's Right Brochure for Sweden'. https://www.tenlaw.uni-bremen.de/Brochures/SwedenBrochure_09052014.pdf.

Baheru, Haymanot. 2017. 'Swedish Legislation of Residential Tenancies : An Interaction between Collective Bargaining and Mandatory Regulation'. *Revista Electrónica de Direito*, no. 3. https://urn.kb.se/resolve?urn=urn:nbn:se:su:diva-148640.

BayObLG, 20-08-1981 – Allg. Reg. 30/81 – Mitteilung von Namen und Anschriften der Vermieter oder Mieter der Vergleichswohnungen. 1981, 1981 NJW 2818–20. BayObLG.

Becker Bütner Held. 2020. 'Vereinbarkeit des „Referenztemperatur-Modells" mit den Vorgaben der Energieeffizienzrichtlinie (EED)'. 2020. https://www.agora-energiewende.de/publikationen/vereinbarkeit-des-referenztemperatur-modells-mit-den-vorgaben-der-energieeffizienzrichtlinie-eed.

Behrendt, Christina. 2002. *At the Margins of the Welfare State: Social Assistance and the Alleviation of Poverty in Germany, Sweden and the United Kingdom.* https://doi.org/10.4324/9781315188034.

Beschlussempfehlung Und Bericht Des Ausschusses Für Recht Und Verbraucherschutz (6. Ausschuss) a) Zu Dem Gesetzentwurf Der Bundesregierung – Drucksachen 19/4672, 19/5415, 19/5647 Nr.13 – Entwurf Eines Gesetzes Zur Ergänzung Der Regelungen Über Die Zulässige Miethöhe Bei Mietbeginn Und Zur Anpassung Der Regelungen Über Die Modernisierung Der Mietsache (Mietrechtsanpassungsgesetz – MietAnpG) 28.11.2018 b) Zu Dem Antrag Der Abgeordneten Caren Lay, Dr. Gesine Lötzsch, Pascal Meiser, Weiterer Abgeordneter Und Der Fraktion DIE LINKE. – Drucksache 19/4829 – Mieterhöhungsstopp Jetzt c) Zu Dem Antrag Der Abgeordneten Caren Lay, Dr. Gesine Lötzsch, Pascal Meiser, Weiterer Abgeordneter Und Der Fraktion DIE LINKE. – Drucksache 19/4885 – Mietenanstieg Stoppen, Mieterinnen Und Mieter Schützen, Verdrängung Verhindern d) Zu Dem Antrag Der Abgeordneten Christian Kühn (Tübingen), Canan Bay- Ram, Daniela Wagner, Weiterer Abgeordneter Und Der Fraktion BÜNDNIS 90/DIE GRÜNEN – Drucksache 19/2976 – Mietrecht Jetzt Wirksam Reformieren – Bezahlbares Wohnen Sichern. 2018. https://dserver.bundestag.de/btd/19/061/1906153.pdf.

BGH, 06.11.2013 – VIII ZR 346/12. 2013. Urt. BGH.

BGH, 10. 4. 2002 – VIII ARZ 3/01 - † Mieterhöhung nach Wärmedämmung ohne Wärmebedarfsberechnung. 2002, 2002 NJW 2036–38. BGH.

BGH, 11-05-1988 – VIII ZR 96/87 – Kein Mitverschulden bei Ansprüchen wegen verspäteter Rückgabe der Mietsache zur Fussnote †. 1988, 1988 NJW-RR 1292–93. BGH.

BGH, 19. 7. 2006 – VIII ZR 212/05 – Erhöhung heizkostenverordnungswidriger Warmmietenabrede. 2006. Urt., 2006 NZM 652–53. BGH.

BGH, 31. 10. 2007 – VIII ZR 261/06 – Wohnflächenbezogene Heizungs- und Warmwasserkostenabrechnung. 2007. Urt., 2008 NJW 142–44. BGH.

Bjarup, Jes. 2005. 'The Philosophy of Scandinavian Legal Realism'. *Ratio Juris* 18 (1): 1–15. https://doi.org/10.1111/j.1467-9337.2005.00282.x.

Blank, Hubert, and Ulf Börstinghaus. 2020. *Miete Kommentar.* 6th ed. München: C. H. Beck.

Bogdan, Michael, ed. 2010. *Swedish Legal System.* Edition 1:1. Institutet För Rättsvetenskaplig Forskning, CXCVII. Stockholm: Norstedts Juridik.

Börstinghaus, Ulf. 2022. *Schmidt-Futterer | Mietrecht.* 15th ed. https://www.beck-shop.de/schmidt-futterer-mietrecht/product/33558884.

Bostadsförvaltningslag (1977:792). n.d. Accessed 24 December 2023. https://www.riksdagen.se/sv/dokument-och-lagar/dokument/svensk-forfattningssamling/bostadsforvaltningslag-1977792_sfs-1977-792/.

Bravo, Giangiacomo, Georgios Pardalis, Krushna Mahapatra, and Brijesh Mainali. 2019. 'Physical vs. Aesthetic Renovations: Learning from Swedish House Owners'. *Buildings* 9 (1): 1–15. https://doi.org/10.3390/buildings9010012.

BSG, 7. 11. 2006 – B 7b AS 16/06 R – Angemessenheit der Unterkunftskosten. 2006. Urt., 2007 NZS 546–50. BSG.

BSG, 26.04.2022 – B 1 KR 5/21 R. 2022. Urt. BSG.

Bub, Wolf-Rüddiger, and Hans-Jörg Kraemer. 2019. *Handbuch der Geschäfts- und Wohnraummiete.* 5th ed. C. H. Beck. https://www.beck-shop.de/bub-treier-handbuch-geschaefts-wohnraummiete/product/16621701.

Buhr, Daniel, Rolf Frankenberger, and Josef Schmid. 2016. 'On the Way to Welfare 4.0? Digitalisation of the Welfare State in Labour Market, Health Care and Innovation Policy: A European Comparison'. https://www.researchgate.net/publication/326147007_On_the_Way_to_Welfare_40_Digitalisation_of_the_Welfare_State_in_Labour_Market_Health_Care_and_Innovation_Policy_A_European_Comparison.

Bundesinstitut für Bau-, Staft- und Raumfoschung. 2016a. 'Mietrecht Und Energetische Sanierung Im Europäischen Vergleich'. 13. https://www.bbsr.bund.de/BBSR/DE/veroeffentlichungen/bbsr-online/2016/bbsr-online-13-2016-dl.pdf?__blob=publicationFile%26v=32016.

Bundesinstitut für Bau-, Staft- und Raumfoschung. 2016b. 'Tenancy Law and Energy Renovation in European Comparison'. 14. http://www.bbsr.bund.de/BBSR/EN/Publications/OnlinePublications/2016/bbsr-online-14-2016-dl.pdf?__blob=publicationFile%26v=4.

Bundesinstitut für Bau-, Staft- und Raumfoschung. 2021. 'Kommunale Wohnungsbestände: Mietengestaltung – Ausweitung – Investitionen. Ergebnisse Der BBSR-Kommunalbefragung 2018'. Bonn. https://www.bbsr.bund.de/BBSR/DE/veroeffentlichungen/sonderveroeffentlichungen/2021/kommunale-wohnungsbestaende-dl.pdf?__blob=publicationFile%26v=2.

Bundesministerium für Wirtschaft und Energie. 2015. 'Energieeffizienzstrategie Gebäude. Wege Zu Einem Nahezu Klimaneutralen Gebäudebestand. Kurzfassung'. Berlin. https://www.bmwi.de/Redaktion/DE/Downloads/E/energieeffizienzstrategie-gebaeude-kurzfassung.pdf?__blob=publicationFile%26v=15.

Bundesregierung. 2020. 'Joint Federal, State and Local Government Housing Strateg. Results of the Housing Summit Held in the Federal Chancellery on 21 September 2018'. https://www.bmi.bund.de/SharedDocs/downloads/EN/themen/building-housing/city-housing/factsheet-german-housing-market.pdf?__blob=publicationFile%26v=2.

Bundesverband Freier Immobilien- und Wohnungsunternehmen. 2018. 'Stellungnahme Zum Gesetzentwurf Der Bundesregierung Entwurf Eines Gesetzes Zur Ergänzung Der Regelungen Über Die Zulässige Miethöhe Bei Mietbeginn Und Zur Anpassung Der Regelungen Über Die Modernisierung Der Mietsache (Mietrechtsanpassunsggesetz – MietAnpG), Bundestag, Drucksache 19/4672'.

Bundesministerium des Innern, für Bau und Heimat. 2019. 'Der Deutschlandatlas, Zu- Und Abwanderung, Datenbasis: Laufende Raumbeobachtung Des BBSR'. Wanderungsstatistik, Regionaldatenbank der Statistischen Ämter des Bundes und der Länder. BMI. https://www.deutschlandatlas.bund.de/DE/Karten/Wo- wir-leben/015-Zuwanderung-Abwanderung.html.

BVerfG, 01.07.1964 – 1 BvR 375/62 – Fiskusprivileg. 1964. Beschl. BVerfG.

BVerfG, 03.12.1969 – 1 BvR 624/56 – Besatzungsschäden, Kriegsfolgeschäden. 1969. Beschl. BVerfG.

BVerfG, 05.06.1973 – 1 BvR 536/72 – Lebach, Lebach-Urteil. 1973. Urt. BVerfG.

BVerfG, 07.06.2005 – 1 BvR 1508/96 – Elternunterhalt. 2005. Urt. BVerfG.

BVerfG, 09.02.2010 – 1 BvL 1/09, 1 BvL 3/09, 1 BvL 4/09 – Hartz IV-Regelsatz, Hartz IV-Gesetz. 2010. Urt. BVerfG.

BVerfG, 11.03.2003 – 1 BvR 426/02 – Benetton-Werbung, Schockwerbung Benetton. 2003. Beschl. BVerfG.

BVerfG, 14.09.1989 – 2 BvR 1062/87 – Verwertung tagebuchartiger Aufzeichnungen, Verwertung tagebuchartiger Aufzeichnungen, Tagebuchaufzeichnung. 1989. Beschl. BVerfG.

BVerfG, 18.06.1975 – 1 BvL 4/74 – Waisenrente. 1975. Beschl. BVerfG.

BVerfG, 18.07.2012 – 1 BvL 10/10, 1 BvL 2/11 –- Asylbewerberleistungsgesetz. 2012. Urt. BVerfG.

BVerfG, 23.07.2014 – 1 BvL 10/12, 1 BvL 12/12, 1 BvR 1691/13 – menschenwürdiges Existenzminimum. 2014. Beschl. BVerfG.

BVerfG, 25.3.2021 – 2 BvF 1/20, 2 BvL 4/20, 2 BvL 5/20 – Nichtigkeit des „Berliner Mietendeckels". 2021. Beschl., 2021 NJW 1377–92. BVerfG.

BVerfG, 29.05.1990 – 1 BvL 20/84, 1 BvL 26/84, 1 BvL 4/86 – Kindergeldentscheidung, Kindergeldkürzung, steuerfreies Existenzminimum. 1990. Beschl. BVerfG.

BVerwG, 14.12.1962 – VII C 140.61. 1962. Urt. BVerwG.

Carlson, Laura. 2012. *The Fundamentals of Swedish Law: A Guide for Foreign Lawyers and Students*. Edition 2. Lund: Studentlitteratur.

Carlson, Laura. 2018. 'The Fundamentals of Swedish Law, Lund 2009.Center for International Climate Research (CICERO): "Second Opinion" on Vasakronan's Green Finance Framework'. Report. CICERO Center for International Climate Research. https://pub.cicero.oslo.no/cicero-xmlui/handle/11250/2720386.

Chen, Jie, and Cecilia Öst. 2005. 'Housing Allowance and the Recipient's Homeownership: Evidence from a Panel Data Study in Sweden'. *Housing Studies* 20 (July). https://doi.org/10.1080/02673030500114540.

CICERO. 2018. '"Second Opinion" on Vasakronan's Green Finance Framework'. https://pub.cicero.oslo.no/cicero-xmlui/bitstream/handle/11250/2720386/vasakronan_2018.pdf?sequence=1%26isAllowed=y.

Commission Staff Working Document Country Report Sweden 2019 Including an In-Depth Review on the Prevention and Correction of Macroeconomic Imbalances Accompanying the Document Communication from the Commission to the European Parliament, the European Council, the Council, the European Central Bank and the Eurogroup 2019 European Semester: Assessment of Progress on Structural Reforms, Prevention and Correction of Macroeconomic Imbalances, and Results of in-Depth Reviews Under Regulation (Eu) No 1176/2011. 2019. https://eur-lex.europa.eu/legal-content/EN/TXT/PDF/?uri=CELEX:52019SC1026&rid=8.

Danelius, Hans. 2015. Mänskliga Rättigheter i Europeisk Praxis : En Kommentar till Europakonventionen Om de Mänskliga Rättigheterna. 5., [Uppdaterade] upp

Council Directive 2003/96/EC of 27 October 2003 Restructuring the Community Framework for the Taxation of Energy Products and Electricity (Text with EEA Relevance). 2003. *OJ L*. Vol. 283. http://data.europa.eu/eli/dir/2003/96/oj/eng.

Derleder, Peter. 2009. 'Die Notwendigkeit Eines Grundrechts Auf Wohnen'. *WuM*, 615–23.

Derleder, Peter. 2013. 'Duldungspflicht Des Mieters Bei Modernisierungsmaßnahmen Mit Energetischem Anspruch'. *Neue Zeitschrift Für Miet- Und Wohnungsrecht*, no. 13: 441–48.

Deutscher Bundestag. 2010. 'Unterrichtung Durch Die Bundesregierung, Energiekonzept Für Eine Umweltschonende, Zuverlässige Und Bezahlbare Energieversorgung Und 10-Punkte-Sofortprogramm – Monitoring Und Zwischenbericht Der Bundesregierung'. Drucksache 17/3049. http://dipbt.bundestag.de/doc/btd/17/030/1703049.pdf.

Deutscher Bundestag. 2012. *Gesetzesentwurf Der Bundesregierung, Entwurf Eines Gesetzes Über Die Energetische Modernisierung von Vermietetem Wohnraum Und Über Die Vereinfachte Durchsetzung von Räumungstiteln (Mietrechtsänderungsgesetz – MietRÄndG)*. http://dip21.bundestag.de/dip21/btd/17/104/1710485.pdf.

Deutscher Bundestag. 2014. *Gesetzesentwurf Der Bundesregierung, Entwurf Eines Gesetzes Zur Dämpfung Des Mietanstiegs Auf Angespannten Wohnungsmärkten Und Zur Stärkung Des Bestellprinzips Bei Der Wohnungsvermittlung (Mietrechtsnovellierugnsgesetz – MietNovG)*. https://www.juris.de/jportal/docs/news_anlage/nlmw/pdf/1803121.pdf.

Deutscher Bundestag. 2018. 'Gesetzentwurf Der Bundesregierung, Entwurf Eines Gesetzes Zur Ergänzung Der Regelungen Über Die Zulässige Miethöhe Bei Mietbeginn Und Zur Anpassung Der Regelungen Über Die Modernisierung Der Mietsache (Mietechtsanpassungsgesetz – MietAnpG)'. Drucksache 19/4672. http://dip21.bundestag.de/dip21/btd/19/046/1904672.pdf≥.

Deutscher Bundestag. 2020. 'Gesetzentwurf Der Fraktionen Der CDU/CSU Und SPD, Entwurf Eines Gesetzes Für Den Erleichterten Zugang Zu Sozialer Sicherung Und Zum Einsatz Und Zur Absicherung Sozialer Dienstleister Aufgrund Des Coronavirus SARS-CoV-2 (Sozialschutz-Paket)'. Drucksache 19/18107. https://dserver.bundestag.de/btd/19/181/1918107.pdf.

Deutscher Bundestag. 2021a. 'Gesetzentwurf Eines Gesetzes Zur Regelung Einer Einmalzahlung Der Grundsicherungssysteme an Erwachsene Leistungsberechtigte Und Zur Verlängerung Des Erleichterten Zugangs Zu Sozialer Sicherung Und Zur Änderung Des Sozialdienstleister-Einsatzgesetzes Aus Anlass Der COVID-19-Pandemie (Sozialschutz.Paket III)'. Drucksache 19/26542. https://dserver.bundestag.de/btd/19/265/1926542.pdf.

Deutscher Bundestag. 2021b. *Gesetzesentwurf Der Bundesregierung, Entwurf Eines Gesetzes Zur Reform Des Mietspiegelrechts (Mietspiegelreformgesetz – MsRG)*. https://dserver.bundestag.de/btd/19/269/1926918.pdf.

Dolzer, Rudolf, Karin Graßhof, and Klaus Vogel. 2004. *Bonner Kommentar Zum Grundgesetz*.

Dürig, Günter, Roman Herzog, and Theodor Maunz, eds. 2021. *Grundgesetz: Kommentar*. München: Beck.

Economidou, Marina. 2014. 'Overcoming the Split Incentive Barrier in the Building Sector'. JRC Publications Repository. 17 September 2014. https://doi.org/10.2790/31513.

Eichenauer, Eva, Jane Irmisch, Peter Ulrich, Ulrike Damerau, Aleksandra Novikova, Jana Karras, and Tim Langenhorst. 2021. 'Regionale Energiewendeszenarien Und Verflechtungsbeziehungen in Berlin Und Dem Landkreis Spree-Neiße'. https://esra-projekt.de/sites/default/files/2021-11/ESRa%20Produkt%204.2_Regionale%20Szenarien.pdf.

'Eigentumsquote'. n.d. Statistisches Bundesamt. Accessed 4 May 2024. https://www.destatis.de/DE/Themen/Gesellschaft-Umwelt/Wohnen/Tabellen/tabelle-eigentumsquote.html.

European Commission. 2013. 'Your Social Security Rights in Sweden. Employment, Social Affairs %26 Inclusion in Sweden'. https://ec.europa.eu/employment_social/empl_portal/SSRinEU/Your%20social%20security%20rights%20in%20Sweden_en.pdf.

Farahani, Abolfazl Sousanabadi. 2017. 'Maintenance, Renovation and Energy Efficiency in the Swedish Multi-Family Housing Market'. Gothenburg: CHALMERS UNIVERSITY OF TECHNOLOGY.

Femenias, Paula, and Anna-Lisa Lindén. 2010. 'Energy Efficiency in the Housing Sector – The Swedish Case'. In .

Ferlin, Maria, and Fryxell Sternbeck. 2020. 'Green Bonds – Big in Sweden and with the Potential Growth'. Sveriges Rijksbank Economic Commentaries No. 12. https://www.riksbank.se/globalassets/media/rapporter/ekonomiska-kommentarer/engelska/2020/green-bonds--big-in-sweden-and-with-the-potential-to-grow.pdf.

Flatow, Beate. 2021. 'Die „ungewollte" Abrechnung: Unwirksame Umlagevereinbarung Oder Inklusivmiete Und § HEIZKV § 2 HeizkostenVO'. *Neue Zeitschrift Für Miet- Und Wohnungsrecht*, no. 23: 865–72.

Fleindl, Hubert. 2016. 'Fleindl: Modernisierung von Wohnraum Und Mieterhöhung'. *Neue Zeitschrift Für Miet- Und Wohnungsrecht*, no. 3: 65–74.

Fritzon, Gustav. 2020. 'Rent Controls. How They Damage the Housing Market, the Economy and Society'. An EPICENTER Publication. http://www.epicenternetwork.eu/wp-content/uploads/2020/01/Epicenter_How-rent-controls-damage-the-housing-market_web.pdf.

Ganal, Helen, Andrea Wiesholzer, Valentina Fongo, Jonas Teitge, and Konstantin Winter. n.d. 'Status Quo Der Energiewenderelevanten Wirtschafts- Und Beschäftigungs- Profile'. https://publica-rest.fraunhofer.de/server/api/core/bitstreams/70fce537-8c06-4c56-a870-88e7cdb21af8/content.

'Gesetz zur Sicherung der Zweckbestimmung von Sozialwohnungen—WoBindG'. n.d. 04 September 2021. https://beck-online.beck.de/Dokument?vpath=bibdata%2Fges%2Fwobindg%2Fcont%2Fwobindg.htm%26anchor=Y-100-G-WOBINDG

Government Offices of Sweden. 2020. 'Memorandum of the Ministry of Health and Social Affairs. Mainstreaming Aging'. https://unece.org/sites/default/files/2021-03/Sweden_CN_EN.pdf.

Grundberg, Johanna. 2016. 'The Right to Adequate Housing for Vulnerable EU Citizens in Sweden | Lund University'. Student publication for professional degree (Master's level). https://www.lunduniversity.lu.se/lup/publication/8874934.

Dreier, Horst. 2015. Grundgesetz-Kommentar 2 von Horst Dreier (Leinen). https://buch-sauermann.de/shop/item/9783161504945/grundgesetz-kommentar-2-von-horst-dreier-leinen.

Gsell, Beate, and Astrid Siegm. 2019. 'Gsell/Siegmund: Mietrecht in Zeiten Des Wohnungsmangels'. *Neue Zeitschrift Für Miet- Und Wohnungsrecht*, no. 14: 489–506.

Gutachterausschuss für Grundstückswerte in den Landkreisen Spree-Neiße und Oberspreewald-Lausitz. 2021. 'Grundstücksmarktbericht 2020'. Cottbus. https://www.gutachterausschuss-bb.de/SPN-OSL/pdf/GMB_SPNOSL.pdf.

Hallof, Irene J. 2013. Das Vermieter-Mieter-Dilemma bei der energetischen Gebäudesanierung: eine rechtliche und ökonomische Analyse. Praxis und Theorie des Bau- und Immobilienrechts 20. Berlin Brüssel: Lexxion Der Juristische Verlag.

Hananel, Ravit, Sharon Krefetz, and Ami Vatury. 2018. 'Public Housing Matters: Public Housing Policy in Sweden, the United States, and Israel'. *Journal of Planning Education and Research* 41 (August): 0739456X1879370. https://doi.org/10.1177/0739456X18793702.

Hansen, Hans, and Marie Louise Schultz-Nielsen. 2015. 'Social Assistance in Five Countries in North-Western Europe'. SSRN Scholarly Paper. Rochester, NY. https://doi.org/10.2139/ssrn.2708360.

Heinz-Herbert, and Stefan Weick. 2014. 'Housing in Germany: Expensive, Comfortable and Usually Rented In'. *ISI – Informationsdienst Soziale Indikatoren*, no. 4. https://www.gesis.org/fileadmin/upload/forschung/publikationen/zeitschriften/isi/isi41-NollWeick-English.pdf.

Henger, Ralph, Sibylle Braungardt, Benjamin Köhle, and Robert Meyer. 2021. 'Wer Zahlt Für Den Klimaschutz Im Gebäudesektor? Reformoptionen Der Modernisierungsumlage. Ariadne Analyse'. Potsdam. https://ariadneprojekt.de/media/2021/07/Ariadne-Analyse_Modernisierungsumlage_August2021.pdf.

Herrlein, Jürgen, and Patrick Tuschl. 2020. 'Herrlein/Tuschl: Das Ringen Um Mietpreisbegrenzung in Bund Und Ländern'. *Neue Zeitschrift Für Miet- Und Wohnungsrecht*, no. 5: 217–35.

Hyresförhandlingslag. n.d. Accessed 24 December 2023. https://www.riksdagen.se/sv/dokument-och-lagar/dokument/svensk-forfattningssamling/hyresforhandlingslag-1978304_sfs-1978-304/.

Hyreskommissionen. 2018. 'Hyresräten i Framtiden'. Stockholm. https://hurvibor.se/wp-content/uploads/hyreskommissionen_rapp.pdf.

Ortwein, Bernard Michael Ortwein. 2003. 'The Swedish Legal System: An Introduction'. Indiana International & Comparative Law Review 13 (2): 405–46. https://doi.org/10.18060/17766.

IKEM, Frauenhofer Institut für Energiewirtschaft und Energiesystemtechnik, Germanwatch, Global Climate Forum, GW Energie, and Leibniz-Institut für Raumbezogene Sozialforschung. 2020. 'ESRa Gesamtvorhabenbeschreibung'.

International Energy Agency. 2019. 'Energy Policies of IEA Countries. Sweden 2019 Review'. IEA. https://www.connaissancedesenergies.org/sites/default/files/pdf-actualites/Energy_Policies_of_IEA_Countries_Sweden_2019_Review.pdf.

Janson, Per. 2002. 'Income and the Swedish Welfare State',. Lund University. https://basicincome.org/bien/pdf/2000Janson.pdf.

Jarass, Hans-D., Martin Kment, and Bodo Pieroth. 2020. *Grundgesetz für die Bundesrepublik Deutschland: Kommentar*. 16. Auflage 2020. München: C.H. Beck.
Jesse, Sven. 2016. 'Instrumente des Energieeffizienzrechts – Analyse und Kriti'. In *Ressourceneffizienz – Leitbild für das Umweltrecht?*, edited by Franz Reimer and Justus-Liebig-Universität Gießen, 1. Auflage, 165–88. Giessener Abhandlungen zum Umweltrecht, Band 26. Baden-Baden: Nomos.
Joint Research Centre of European Commission (JRC). 2017. 'Overcoming the Split Incentive Barrier in the Building Sector. Unlocking the Energy Efficiency Potential in the Rental & Multifamily Sectors'. http://publications.jrc.ec.europa.eu/repository/bitstream/JRC101251/ldna28058enn.pdf.
Jordabalk (1970:994) (JB) | Lagen.Nu. n.d. Accessed 24 December 2023. https://lagen.nu/1970:994.
'Just over 5 Million Dwellings in Sweden'. n.d. Statistiska Centralbyrån. Accessed 24 December 2023. https://www.scb.se/en/finding-statistics/statistics-by-subject-area/housing-construction-and-building/housing-construction-and-conversion/dwelling-stock/pong/statistical-news/dwelling-stock-2020-12-31/.
Kanalan, Ibrahim. 2018. 'Kanalan: Rechtstreue Als Voraussetzung Für Den Anspruch Auf Die Gewährleistung Eines Menschenwürdigen Existenzminimums'. *Neue Zeitschrift Für Sozialrecht*, no. 16: 641–46.
Karras, Jana, and Aleksandra Novikova. 2021. 'Germany. Legal Barriers to the Decarbonisation of Dwellings Occupied by Low-Income Tenants and Financial Incentives to Overcome These Barriers'. IKEM Climate Finance Week. https://www.ikem.de/wp-content/uploads/2021/05/Brief_5_Legal-barriers-to-the-decarbonisation-of-dwellings.pdf.
Kaßler, Martin. 2020. 'Kaßler: Gebäudeenergiegesetz: Gute Ansätze, Aber Rechtsunsicherheit Für Immobilienverwalter Bleibt Bestehen'. *Zeitschrift Für Wohnungseigentumsrecht*, no. 1: 58–60.
Katz, Alfred, and Gerald G. Sander. 2019. *Staatsrecht Staatsorganisation und Grundrechte*. 19th ed. Heidelberg: C. F. Mueller Verlag GmbH.
Kischel, Uwe. 2015. *Rechtsvergleichung*. München: C.H. Beck.
Klinski, Stefan, Ass Friedhelm Keimeyer, and Sibylle Braungardt. 2021. 'Teilwarmmietemodelle Im Wohnungsmietrecht Als Geeignetes Instrument Zum Klimaschutz? Kurzstudie Zur Rechtlichen Und Praktischen Machbarkeit'. Öko-Institut e.V. https://www.oeko.de/fileadmin/oekodoc/SozUP-Teilwarmmieten.pdf.
'Knapp ein Viertel der Haushalte heizt die Wohnung 2018 immer noch mit Öl'. n.d. Statistisches Bundesamt. Accessed 23 December 2023. https://www.destatis.de/DE/Presse/Pressemitteilungen/2019/10/PD19_N002_129.html.
Knauff, Matthias. 2022. 'Stand Und Perspektiven Des Landesklimaschutzrechts'. *Klima Und Recht*, no. 2: 47–52.
Knutagård, Marcus, Josephine Heap, and Kenneth Nelson. 2019. *ESPN Thematic Report on National Strategies to Fight Homelessness and Housing Exclusion – Sweden*.
Lag om ekonomiska föreningar. n.d. Accessed 24 December 2023. https://www.riksdagen.se/sv/dokument-och-lagar/dokument/svensk-forfattningssamling/lag-2018672-om-ekonomiska-foreningar_sfs-2018-672/.
Lag om kooperativ hyresrätt. n.d. Accessed 24 December 2023. https://www.riksdagen.se/sv/dokument-och-lagar/dokument/svensk-forfattningssamling/lag-200293-om-kooperativ-hyresratt_sfs-2002-93/.
Lindner, Eric. 2019. *Wohnraummietrecht*. 3. Auflage. NJW-Praxis 19. München: C.H. Beck.
Lorenz, Stephan, and Christoph Reymann. n.d. 'Beck-Online.Großkommentar Zum Zivilrecht'. Accessed 24 November 2023. https://beck-online.beck.de/Dokument?vpath=bibdata%2Fkomm%2FBeckOGK%2Fcont%2FBeckOGK.Inhaltsverzeichnis.htm&anchor=Y-400-W-

BECK-ONLINE.GROSSKOMMENTAR%26jumpType=Jump%26jumpWords=beck-online.GROSSKOMMENTAR%2Bzum%2BZivilrecht%26readable=Suche%2Bnach%2BWerk%253a%2Bbeck-online.GROSSKOMMENTAR%2Bnur%2Bin%2Bmeinen%2BModulen.

Lunderup, Erik. 2016. 'Kooperativ Hyresrätt En Användbar Upplåtelseform För Ökad Nyproduktion i Sverige'. Lund. https://lup.lub.lu.se/luur/download?func=downloadFile%26recordOId=8881958%26fileOId=8881961.

Mann, Thomas, Christoph Sennekamp, Michael Uechtritz, Pino Bosesky, Ines Härtel, Foroud Shirvani, and Thomas Smollich. 2019. *Verwaltungsverfahrensgesetz: Großkommentar*. 2. Auflage. NomosKommentar. Baden-Baden: Nomos.

Manthey, Eva. 2021. 'Die Innovationsklausel Des Gebäudeenergiegesetzes'. *Zeitschrift Für Das Gesamte Recht Der Energiewirtschaft*, no. 5: 147–51.

Maunz, Theodor, and Günter Dürig. 2021. *Grundgesetz Kommentar*. 94. Lieferung. München.

Ministry of the Environment and Energy of Sweden. 2017. 'The Swedish Climate Policy Framework'. https://www.government.se/495f60/contentassets/883ae8e123bc4e42aa8d59296ebe0478/the-swedish-climate-policy-framework.pdf?TSPD_101_R0=082953afa5ab2000860fac6cc18d892e8b02c06e1dd9048c5134977e9b670fe0ec1830a07a743ef70823abe9d1143000af12e03a5fa4764c8bc9d946cd010617b06992fc5eb0d10ff214a763fb94129ccc49ddb50cc517b486650aa34822c797.

Möbert, Jochen. 2021. 'Outlook for the German Residential Property Market 2021 and beyond. House Price Cycle Could End in 2024'. Deutsche Bank Research. https://www.dbresearch.com/PROD/RPS_EN-PROD/PROD0000000000517463/Outlook_for_the_German_residential_property_market.pdf?undefined%26realload=7HoNcHdkSSFOhVUGzgFtjP~r77OOQp-4tEXuIIrVdg12Zh9WfyxJSwAGiwbHoXuNe.

'MuekoBGB – Beck-Online'. n.d. Accessed 23 December 2023. https://beck-online.beck.de/Dokument?vpath=bibdata%2Fkomm%2FMuekoBGB%2Fcont%2FMuekoBGB.htm%26anchor=Y-400-W-MUEKOBGB%26jumpType=Jump%26jumpWords=M%25c3%25bcKoBGB%26readable=Suche%2Bnach%2BKommentar%253a%2BM%2526%2523252%253bkoBGB.

Münch, Ingo von, Philip Kunig, Jörn Axel Kämmerer, and Markus Kotzur, eds. 2021. *Grundgesetz Kommentar*. 7., Neubearbeitete Auflage. München: C.H. Beck.

Naturvardsverket. 2021. 'Report of Sweden on Climate Policies and Measures and on Projections'. https://www.naturvardsverket.se/upload/miljoarbete-i-samhallet/uppdelat-efter-omrade/klimat/prognoser-for-Sveriges-utslapp/report-sweden-march-2021.pdf.

Nelson, Kenneth, and Johan Fritzell. 2016. 'ESPN Thematic Report on Minimum Income Schemes Sweden'. European Commision.

Nilsson, J., and A. Mårtensson. 2003. 'Municipal Energy-Planning and Development of Local Energy-Systems'. *Applied Energy* 76 (September): 179–87. https://doi.org/10.1016/S0306-2619(03)00062-X.

Norberg, Per, and Jakob Juul-Sandberg. 2016. 'Rent Control and Other Aspects of Tenancy Law in Sweden, Denmark and Finland: The European Network for Housing Research Conference'. In . http://www.enhr2016.com/abstracts/.

Novikova, Aleksandra, Tim Langenhorst, Jana Karras, Peter Ulrich, and Ulrike Damerau. n.d. 'Synopsis Der Handlungsoptionen, Energiewende Im Sozialen Raum'. Produkt No. 4.1. https://esra-projekt.de/sites/default/files/2021-08/esra_produkt_4.1_statusquo_allianzen_und_finanzierung_.pdf.

OECD. 2017. 'Sweden Country Review'. https://www.oecd.org/env/country-reviews/2451763.pdf.

OECD. 2019. 'Economic Surveys: Sweden'. https://www.oecd.org/economy/surveys/OECD-economic-surveys-sweden-2019-overview.pdf.

OECD. 2020. 'The OECD Tax-Benefit Model for Sweden. Description of Policy Rules'. https://www.oecd.org/els/soc/TaxBEN-Sweden-2020.pdf.

Om Boverkets Byggregler, BBR – Boverket. n.d. Accessed 24 December 2023. https://www.boverket.se/sv/byggande/regler-for-byggande/om-boverkets--byggregler-bbr/#.

Persson, Johannes, and Stefan Grönkvist. 2015. 'Drivers for and Barriers to Low-Energy Buildings in Sweden'. *Journal of Cleaner Production*, Special Issue: Toward a Regenerative Sustainability Paradigm for the Built Environment: from vision to reality, 109 (December): 296–304. https://doi.org/10.1016/j.jclepro.2014.09.094.

Pierre, Jon, Sven Jochem, and Detlef Jahn. 2016. 'Sweden Report. Sustainable Governance Indicators 2016'. https://www.sgi-network.org/docs/2016/country/SGI2016_Sweden.pdf.

Radbruch, Gustav. 1950. *Rechtsphilosophie*. 4. Vol. I. Stuttgart.

Regerinskansliet. 2013. 'The Constitution of Sweden'. Stockholm. https://www.government.se/contentassets/7b69df55e58147638f19bfdfb0984f97/the-constitution-of-sweden/.

Ring, Gerhard, and Line Olsen-Ring. 2014. *Einführung in das skandinavische Recht*. 2. Aufl. Schriftenreihe der Juristischen Schulung Ausländisches Recht 137. München: Beck.

Rolf, Hannes. 2021. 'A Union for Tenants: Tenant Militancy in Gothenburg as a Historical Example'. *Radical Housing Journal* 3 (1): 167–86.

Rothstein, Bo. 2010. *Vad bör staten göra? : om välfärdsstatens moraliska och politiska logik*. SNS (Studieförb. Näringsliv och samhälle. https://urn.kb.se/resolve?urn=urn:nbn:se:uu:diva-413050.

Ruhland, Johannes, Christian Alexander Kreibich, and Ralf Herud. n.d. 'Ökonomische Analyse Der Modernisierungsanreize Im Deutschen Mietrecht'. *Jena Research Papers in Business and Economics* 04 (2010). https://oweb.b67.uni-jena.de/Papers/wp-jbe201004.pdf.

Säcker, Franz Jürgen, Roland Rixecker, Hartmut Oetker, and Bettina Lipmerg, eds. n.d. MuekoBGB - Beck-Online. Accessed 23 December 2023. https://beck-online.beck.de/Dokument?vpath=bibdata%2Fkomm%2FMuekoBGB%2Fcont%2FMuekoBGB.htm&anchor=Y-400-W-MUEKOBGB&jumpType=Jump&jumpWords=M%25c3%25bcKoBGB&readable=Suche%2Bnach%2BKommentar%253a%2BM%2526%2523252%253bkoBGB.

Schach, Klaus, Michael Schultz, and Peter Schüller, eds. n.d. BeckOK MietR | Inhaltsübersicht - Beck-Online. Accessed 4 May 2024. https://beck-online.beck.de/Dokument?vpath=bibdata%2Fkomm%2FBeckOKMietR_35%2Fcont%2FBeckOKMietR.Inhaltsverzeichnis.htm&anchor=Y-400-W-BECKOKMIETR&jumpType=Jump&jumpWords=BeckOK%2BMietrecht&readable=Suche%2Bnach%2BKommentar%253a%2BBeckOK%2BMietrecht.

Schmitt, Carl. 1928. *Verfassungslehre*. Leipzig/München.

Schoch, Friedrich, and Jens-Peter Schneider. 2021. *Verwaltungsrecht*. München: C. H. Beck.

SG Aachen, 20.07.2011 – S 5 AS 475/11. 2011. Urt. SG Aachen.

Smits, J. M., and Edward Elgar Publishing, eds. 2006. *Elgar Encyclopedia of Comparative Law*. Elgar Original Reference. Cheltenham, UK ; Northampton, Mass: Edward Elgar Pub.

Socialförsäkringsbalk (2010:110). n.d. Accessed 24 December 2023. https://www.riksdagen.se/sv/dokument-och-lagar/dokument/svensk-forfattningssamling/socialforsakringsbalk-2010110_sfs-2010-110/.

Socialtjänstlag (2001:453). n.d. Accessed 24 December 2023. https://www.riksdagen.se/sv/dokument-och-lagar/dokument/svensk-forfattningssamling/socialtjanstlag-2001453_sfs-2001-453/.

Sodan, Helge, and Jan Ziekow. 2018. *Verwaltungsgerichtsordnung: Großkommentar*. 5. Auflage. NomosKommentar. Baden-Baden: Nomos.

Spitzlei, Thomas. 2017. 'Asylbewerberleistungsrecht Und Grundgesetzliches Existenzminimum'. *Juristische Arbeitsblätter*, no. 3: 165–71.

Statens medicinsk-etiska råd. 2012. 'Männisovärdet i Svensk Tillämping, in: Statens Medicinsk-Etiska Råd: Det Svårfångade Människovärdet – En Debattskrift, Omarbetad Upplaga'. Stockholm.

Statistisches Bundesamt. 2021. 'Statistik: Eigentümerquote Nach Bundesländern'. 16 June 2021. https://www.destatis.de/DE/Themen/Gesellschaft-Umwelt/Wohnen/Tabellen/eigentuemer-quote-nach-bundeslaender.html.

Stede, Jan, Franziska Schütze, and Johanna Wietschel. n.d. 'DIW Berlin: Wärmemonitor 2019: Klimaziele bei Wohngebäuden trotz sinkender CO_2-Emissionen derzeit außer Reichweite'. Text DIW Wochenbericht Nr. 40/2020. DIW Berlin. Accessed 2 December 2023. https://doi.org/10.18723/diw_wb:2020-40-1.

Stromholm, Stig. 2013. An Introduction to Swedish Law: Volume 1. Softcover reprint of the original 1st ed. 1981 Edition. Dordrecht: Springer.

Svensk Författningssamling. n.d. http://rkrattsdb.gov.se/SFSdoc/03/030593.PDF.

'Sweden: Housing Allowance Recipients by Type 2021'. n.d. Statista. Accessed 24 December 2023. https://www.statista.com/statistics/536899/sweden-housing-allowance-recipients-by-type-of-household/.

'Sweden Population (2023) – Worldometer'. n.d. Accessed 24 December 2023. https://www.worldometers.info/world-population/sweden-population/.

Swedish Climate Policy Council. 2019. 'Report of the Swedish Climate Policy Council'. Stockholm. https://www.klimatpolitiskaradet.se/wp-content/uploads/2019/09/climatepolicycouncilreport2.pdf.

Swedish Environmental Protection Agency. 2017. 'Swedish Environmental Law. An Introduction to the Swedish System for Environmental Protection'. 6790. https://www.naturvardsverket.se/Documents/publikationer6400/978-91-620-6790-8.pdf?pid=21184.

Swedish Ministry of Environment and Energy. 2014. 'Sweden's Draft Integrated National Energy and Climate Plan. According to Regulation (EU) 2018/1999 of the European Parliament and of the Council on the Governance of the Energy Union and Climate Action'. Text. Regeringskansliet. Regeringen och Regeringskansliet. 24 September 2014. https://www.government.se/4a9ef2/contentassets/e731726022cd4e0b8ffa0f8229893115/swedens-draft-integrated-national-energy-and-climate-plan.

Swiss Institute of Comparative Law. 2016. 'Gutachten Zum Recht Der Energetischen Sanierung Und Wertvermehrenden Verbesserung Im Mietrecht in Deutschland, Frankreich, Italien, Österreich, Schweden, Spanien Und Im Vereinigten Königreich (England)',. Lausanne. https://www.google.com/url?sa=t%26rct=j%26q=%26esrc=s%26source=web%26cd=%26cad=rja%26uact=8%26ved=2ahUKEwjP0bGJg-qCAxUphv0HHZ6sAN4QFnoECBgQAQ%26url=https%3A%2F%2Fwww.bwo.admin.ch%2Fdam%2Fbwo%2Fde%2Fdokumente%2F04_Mietrecht%2F46_Studien_und_Publikationen%2Frechtsvergleiche%2FSIR-Energetische_Sanierung.pdf.download.pdf%2FSIR-Energetische_Sanierung.pdf%26usg=AOvVaw2Z97KkvAtwHjhlLgcO6yK7%26opi=89978449.

Terner Center for Housing Innovation. 2017. 'Housing In Sweden: An Overview'. https://ternercenter.berkeley.edu/research-and-policy/housing-in-sweden-an-overview/.

The Instrument of Government. n.d. Accessed 24 December 2023. https://www.riksdagen.se/globalassets/07.-dokument--lagar/the-instrument-of-government-2015.pdf.

Thema, Johannes, and Florin Vondung. 2020. 'EPOV Indicator Dashboard. Methodology Guidebook'. Wuppertal Institut für Klima, Umwelt, Energie GmbH. https://energy-poverty.ec.europa.eu/system/files/2021-09/epov_methodology_guidebook_1.pdf.

Theobald, Christian, and Jürgen Kühling. 2021. *Energierecht*. München: C. H. Beck.

Thomaßen, Georg, Leo Reutter, Alexandra Langenheld, and Matthias Deutsch. 2020. 'Wie passen Mieterschutz und Klimaschutz unter einen Hut?' 190/06-I-2020/DE. Agora Energiewende.

https://static.agora-energiewende.de/fileadmin/Projekte/2020/2020_07_Warmmieten/A-EW_190_Mieterschutz_Klimaschutz_WEB.pdf.

Thomaßen, Georg, Leo Reutter, Alexandra Langenheld, and Matthias Deutsch. 2021. 'CO2 Emissions Trading in Buildings and the Landlord-Tenant Dilemma: How to Solve It. A Proposal to Adjust the EU Energy Efficiency Directive'. Agora Energiewende. https://static.agora-energiewende.de/fileadmin/Projekte/2020/2020_07_Warmmieten/A-EW_207_Emissions-Trading-Buildings-Landlord-Tenant-Dilemma_WEB.pdf.

Union, Publications Office of the European. 2019. 'Commission Staff Working Document the Eu Environmental Implementation Review 2019 Country Report – Sweden Accompanying the Document Communication from the Commission to the European Parliament, the Council, the European Economic and Social Committee and the Committee of the Regions the Eu Environmental Implementation Review 2019: A Europe That Protects Its Citizens and Enhances Their Quality of Life, Swd/2019/117 Final/2'. Website. Publications Office of the EU. Publications Office of the European Union. 31 July 2019. https://op.europa.eu/en/publication-detail/-/publication/51b6edfe-06ed-11ea-8c1f-01aa75ed71a1/language-en/format-PDF/source-261568670.

Vasakronan. 2018. 'VASAKRONAN GREEN FINANCE FRAMEWORK'. https://wwwvasakronanse.cdn.triggerfish.cloud/uploads/2018/10/vasakronan_green_bond_framework.pdf.

Verband des Zimmerer- und Holzbaugewerbes Baden-Württemberg. 2019. 'Anmerkung Zum Referenten-Entwurf Des „Gesetzes Zur Einsparung von Energie Und Zur Nutzung Erneuerbarer Energie Zur Wärme- Und Kälteerzeugung in Gebäuden"'. https://www.bmwi.de/Redaktion/DE/Downloads/Stellungnahmen/Stellungnahmen-GEG/Verbaende/holzbau.pdf?__blob=publicationFile%26v=6.

VG Greifswald, 22.09.2020 – 2 A 428/20 HGW. 2020. Urt. VG Greifswald.

VGH München, 11. 5. 2006 – 8 ZB 485/06 – Verkehrssicherungspflicht der Wohnungseigentümer für „Trampelpfad". 2006, 2006 NZM 595. VGH München.

Wiesholzer, Andrea, Jasmin Koller, Peter Ulrich, Janne Irmisch, and Eva Eichenauer. 2021. Regionale Disparitäten Und Einstellungen Zur Energiewende. Berlin Und Spree-Neiße Im Vergleich.

Wilhelmsson, Mats. 2021. 'Demand for Rent-Regulated Apartments in the Swedish Housing Market'. KTH Royal Institute of Technology.

Wissenschaftliche Dienste des Deutschen Bundestages. 2019. 'Sachstand: Recht Auf Wohnen. Ausgestaltung Und Rechtswirkung in Den Verfassungen Der Bundesländer Und Der EU-Mitgliedsstaaten'. WD3 – 3000 – 120/19. https://www.bundestag.de/resource/blob/651544/50f6cb8ef28a8b472f0fa00add53d78a/WD-3-120-19-pdf-data.pdf.

'Wohnungsbestand Ende 2019: 42,5 Millionen Wohnungen'. n.d. Statistisches Bundesamt. Accessed 23 December 2023. https://www.destatis.de/DE/Presse/Pressemitteilungen/2020/07/PD20_281_31231.html.

6 Possible Solutions

6.1 Introduction

As identified in the previous chapter, both Sweden and Germany encounter legal challenges in the decarbonisation of rented dwellings occupied by low-income tenants. Employing the contextual method of comparative law in this research allows for the examination of not only the rental law context but also the non-legal factors that may offer solutions to the landlord-tenant dilemma. Accordingly, the fourth research question, as formulated in Sect. 5.1, is allocated a separate chapter for focused analysis. This question aims to explore alternative climate financing instruments suitable for the decarbonisation of dwellings occupied by low-income households. The subsequent step involves comparing the financial instruments used in Germany and Sweden. If Sweden employs successful instruments not yet adopted in Germany, Chap. 7 will assess their replicability in the German context.

The analysis will primarily focus on instruments that mobilise private investments for several reasons. First, the substantial investment required for the energy transition cannot solely rely on the public sector, particularly in economically strained regions or those with limited public budgets.[1] The EGD has emphasised the importance of private investments in this regard.[2] Second, employing public financial instruments often necessitates legislative amendments, a process that is generally time-consuming and potentially fraught with constitutional risks, such as tax reforms or the introduction of special

[1] Karras/Novikova 2021, p. 26.

[2] COM(2019) 640 final.

J. Karras, *Legal Barriers to the Energy Modernisation of Dwellings Occupied by Low-income Tenants and Opportunities to Overcome these Barriers*, Schriftenreihe des Instituts für Klimaschutz, Energie und Mobilität,
https://doi.org/10.1007/978-3-658-44193-7_6

levies.[3] While possible public financial instruments for the energy modernisation of housing occupied by low-income households will be briefly outlined, they will not be the primary focus of the detailed discussion.

Within one of the working packages,[4] of the project "Energy Transition in Social Space" (Energiewende im sozialen Raum—ESRa)[5] the IKEM conducted legal and economic research to identify available and potential financial instruments for energy transition in Berlin and the Spree-Neiße district. This involved a literature review and a legal framework screening to ascertain the status quo and best practices from other states, German *Länder*, and municipalities regarding financial instruments for energy transition. Hypotheses on potential financing options were formulated based on the literature review, categorising them into public and private financial instruments. Interviews with stakeholders in energy transition (including researchers, public authorities, and financial intermediaries) were conducted to evaluate these hypotheses. Relevant public financial instruments for addressing the landlord-tenant dilemma include tax and/or special levy reforms, and financing through CO_2 pricing under the Federal Emissions Trading Act (BEHG).[6] The advantages and disadvantages of these instruments were discussed in stakeholder interviews. The outcomes of this working package will inform the evaluation of these and additional instruments (such as KfW programmes and partial warm rent) in the context of decarbonising rented dwellings, thereby addressing the fifth research question of the comparative analysis.

6.2 Are Any Instruments Outside Rental Law Applicable to Decarbonising Dwellings Occupied by Low-Income Tenants?

6.2.1 Germany

6.2.1.1 KfW Programmes

The Kreditanstalt für Wiederaufbau (KfW) is established as a public law institution under §1 (1) s. 1 of the Kreditanstalt für Wiederaufbau Act (KredAnstWiAG). According to §1 (2) KredAnstWiAG, the total share capital of KfW is three billion, seven hundred and fifty million euros, with the Federal Government holding three billion euros and the *Länder* (federal states) holding seven hundred and fifty million euros. The main tasks of the KfW, as outlined in §2 (1) KredAnstWiAG, include implementing support

[3] See more: BVerfGE 93, 319 (342); BVerfG, 3.2.2009 – 2 BvL 54/06; ESRa, P4.3, p. 20 ff.

[4] ESRa P4.3. .

[5] https://esra-projekt.de..

[6] See more: ESRa, P4.3.

measures, particularly financing, on behalf of the state in specific areas detailed in this legal norm. The KfW is authorised to provide loans and other financing forms to regional authorities and special-purpose associations under public law. Additionally, it can finance projects with purely social objectives, education-promoting measures, and other initiatives that serve the interests of the German and European economies, as per §2 (1) KredAnstWiAG. The KfW operates as a national development bank with a unique business model, distinguishing it from credit institutions in the private, cooperative, or public sectors. This distinction primarily arises from the regulation under §2 (1) No. 2 of the German Banking Act (KWG). According to this legal provision, KfW is not considered a credit institution within the German banking system. It is only subject to the regulations of the KWG to a limited extent, as stated in §2 (2) KWG.[7]

As a development bank, the KfW offers several programmes related to the energy modernisation of residential buildings. These programmes provide loans and grants for energy-efficient refurbishments and the replacement of electricity and heating systems with more energy-efficient alternatives.[8] Funding is available for structural and technical measures that enhance energy efficiency and meet KfW's minimum requirements.[9]

However, these financial instruments are only available as additional financing to the existing financial assets of the applicants.[10] This structure does not directly resolve the landlord-tenant dilemma, particularly for low-income tenants who lack the necessary financial resources to qualify. Furthermore, landlords, despite the availability of grants, often show limited interest in investing in energy modernisation, as they do not directly benefit from the resultant energy savings. This situation has been outlined in previous chapters.

Consequently, the current mechanisms offered by the KfW are not effective in facilitating the decarbonisation of residential buildings inhabited by low-income tenants. These mechanisms fail to incentivise landlords sufficiently and are inaccessible to tenants who do not possess the required financial means. Therefore, the reach of the KfW programmes and impact in addressing the specific challenges faced by low-income tenants in Germany's rental housing sector are limited.

[7] BBundesrat Drucksache 215/13, p. 1.

[8] 'Förderprodukte für energieeffiziente Sanierung – Übersicht | KfW'. n.d. Accessed 24 December 2023. https://www.kfw.de/inlandsfoerderung/Privatpersonen/Bestandsimmobilie/Energieeffizient-Sanieren/Förderprodukte/. .

[9] See more: KfW. 2020. 'Anlage Zu Den Merkblättern Energieeffizient Sanieren - Kredit Und Investitionszuschuss'. 151/152/430. https://www.kfw.de/PDF/Download-Center/Förderprogramme-(Inlandsförderung)/PDF-Dokumente/6000003612_M_151_152_430_Anlage_TMA_2018_04.PDF. .

[10] 'Förderprodukte für energieeffiziente Sanierung – Übersicht | KfW'. n.d. Accessed 24 December 2023. https://www.kfw.de/inlandsfoerderung/Privatpersonen/Bestandsimmobilie/Energieeffizient-Sanieren/Förderprodukte/. .

6.2.1.2 Reform of Public Law Levies

Environmental economists and legal experts have extensively debated the use of tax law instruments to discourage environmentally harmful practices. One of the key objectives is to internalise the costs associated with environmental damage, particularly CO_2 emissions.[11] Various approaches have been proposed to achieve this internalisation.

One such approach is the interventionist model, which mandates ex-ante cost allocation. This model is characterised by political regulations that specify the burdens to be imposed on polluters before the occurrence of external effects. It involves predetermined rules that define how polluters are to be held accountable for their environmental impact.[12] Another widely discussed approach, especially in the context of decarbonising the residential building sector, is a regulatory solution, exemplified by the use of emission certificates. This system organises social relations in a way that requires polluters to compensate for their environmental impact without intervention from external entities. The internalisation of initially external costs and benefits occurs "automatically" through a market organised competitively.[13] Polluters are given the choice to either modify their behaviour or pay a monetary sum to offset the negative effects of their actions. The market mechanism effectively transforms environmental protection into an economically rational choice, encouraging polluters to adopt more sustainable practices or bear the financial costs of their environmental impact.[14] This market-based approach incentivises investment in environmentally friendly technologies, as they become more cost-effective for polluters.[15]

CO_2-Pricing through ETS

On 20.12.2019 the "Act on National Certificate Trading for Fuel Emissions" (Federal Fuel Emissions Trading Act, BEHG)[16] entered according to its §24 (1) into force following the promulgation in the Federal Law Gazette (Bundesgesetzblatt). This Act is part of the so-called "climate packages" aiming to introduce a series of interrelated measures to achieve the European climate targets by 2030 and the long-term goal of climate neutrality by 2050.[17] According to §1 s. 2 of the BEHG, the purpose of this Act is to price fossil greenhouse gas emissions including CO_2. To implement this CO_2 pricing, a national certificate trade for fuel emissions has been established, the rationale being to transform the

[11] Klinski/Keimeyer 2019, p. 1465 f.; Faure 2020, p. 142ff.; Rodi, 2016, p. 531.

[12] Rodi 2014, p. 39.

[13] Rodi 2014, p. 39.

[14] Klinski/Keimeyer 2019, pp. 1465 f.

[15] Ibid.

[16] BGBGl. I 2019, p. 2728.

[17] Wernsmann 2020, p. 3.

emission of greenhouse gases into a chargeable commodity by linking emissions to the possession of certificates.[18]

The national emissions trading system, as regulated by the BEHG, covers emissions resulting from the combustion of fossil fuels. This includes heating oil, liquefied petroleum gas, natural gas, coal, petrol, and diesel, as well as emissions produced by the heating of buildings in the building sector.[19] The scope of application of the BEHG essentially mirrors that of the Energy Tax Act (EnergieStG) for fossil fuels. In this context, §2 (2) S. 1 BEHG refers to the qualifying events of § 8 (1) S. 1 of the Energy Tax Act. Consequently, the act of introducing fuels into the market is a determining factor for participation in the emission certificate trading. The subsequent use of these fuels by end consumers in private households is not relevant.[20]

The enactment of the BEHG has been aligned in line with the German Governmental Climate Protection Programme 2030.[21] The aim of CO_2 pricing is to encourage investment in more efficient or entirely climate-neutral technologies, thereby making them more economical. In this respect, integrating the building sector into the Emission Trading System (ETS), whether European or national, could facilitate the reduction of CO_2 emissions through the limited number of emission certificates. However, due to the specific non-legal context in Germany, CO_2 pricing cannot fully exert its intended effect in the building sector, as over half of the households in Germany are rentals. CO_2 pricing can incentivise owners of residential properties to invest in energy modernisation. Yet, in the case of rented housing, the landlord-tenant dilemma resurfaces, undermining the incentive effect of CO_2 pricing. The landlords are not directly impacted by rising energy costs, as these are borne by the tenants, resulting in a lack of motivation to invest in retrofitting the properties. On the other hand, tenants, while bearing the brunt of price increases, have limited means—apart from changing their energy consumption behaviour—to influence emission reduction.[22] In the large cities like Berlin, where the share of rented dwellings exceeds 80%,[23] the inefficiency of CO_2 pricing is even more pronounced.

Currently, the BEHG does not integrate the building sector into the emission trading system. It incorporates various regulatory elements derived from the EU ETS. However,

---

[18] Ibid, p. 4.

[19] Bundestag, Drucksache 19/14746, p. 2; Kreuter-Kirchhof 2022, p. 73.

[20] Wernsmann/Bering 2020, p. 498.

[21] Bundesregierung 2019; Bundestag, Drucksache 19/25246, p. 1.

[22] Bundestag Drucksache 19/25246, p. 1f.

[23] Karras/Novikova 2021, p. 26; 'Wer profitiert vom Berliner Mietmarkt?' 2019. Tagesspiegel. 28 June 2019. https://interaktiv.tagesspiegel.de/lab/mieten-und-renditen/. .

there are key differences between these two systems. Firstly, the BEHG establishes an "upstream emission trading", in contrast to the "downstream emission trading" system of the EU ETS. Secondly, the German ETS sets an absolute cap on available emission allowances, starting from 2026. The first five years since the enactment of the BEHG, along with a transition year in 2016, feature a fixed price corridor. From 2027 onwards, the certificate market will operate with free pricing. Until 2026, there is a cap, as outlined in §10 BEHG, which can be exceeded under §§4 and 5 BEHG.[24]

The so-called upstream emissions trading can be regarded as a specific form of allowance trading. In this model, it is the companies at an upstream economic level, those responsible for bringing emissions-intensive products to the market, that are obliged to procure the necessary certificates, rather than the actual emitters. The cost of these certificates is then typically passed on to the end consumers. These consumers are subsequently tasked with adopting measures to reduce emissions. However, in contrast to the downstream model, companies obligated to participate in emissions trading in the upstream model have little incentive to undertake measures to reduce emissions. Consequently, involving the building sector in the emissions trading system would be more practical using this upstream model. Otherwise, every individual consumer of fuels, such as heating oil or natural gas, would be required to purchase and, if necessary, trade certificates—a scenario that would be unfeasible due to the sheer volume of transactions involved.[25]

CO_2 pricing could become an effective tool for reducing emissions in owner-occupied housing, as it creates additional incentives for owners to decarbonise their buildings through energy modernisation. However, in the rental housing sector, the landlord-tenant dilemma undermines the effectiveness of CO_2 pricing. Here, the tenants, who bear the brunt of the price increases, have limited ability to influence emission reductions, being restricted mostly to energy-saving behavioural changes. In contrast, landlords, who control the building and its technical installations, are not directly affected by rising prices and thus lack the incentive to invest in decarbonisation measures.[26]

Additionally, there are constitutionally-based legal uncertainties regarding the national emissions trading system. The obligation for certain companies to purchase emission allowances translates into a monetary payment to the state for the fuels they sell to third parties, effectively compensating financially for the CO_2 emissions produced when these fuels are combusted. From a constitutional perspective, this monetary payment could be legally interpreted as a special levy, as it contributes to a public budget. The classification of this payment, whether as a tax ("*steuerliche Abgabe*") or a non-tax

[24] Klinski/Keimeyer 2020, p. 343.

[25] Ibid, p. 343.

[26] Bundestag, Drucksache 19/25246, p. 2.

levy ("*nicht steuerliche Abgabe*"), has significant implications. It affects not only the admissibility requirements but also raises questions regarding legislative competence.[27]

The BVerfG has developed significant case law regarding the taxes. The Federal Republic is a "tax state" (*Steuerstaat*), implying that it must primarily finance its tasks through tax revenue.[28] This principle is derived from the Grundgesetz (GG), which explicitly identifies taxes as the sole sources of revenue in Articles 105 and 106 GG. Consequently, levies to be paid to the state that are not classified as taxes are only permissible under specific conditions. Regarding special levies that do not contradict the GG, the BVerfG has developed case groups for non-tax levies.[29] While the state is restricted to levying only the existing taxes, constitutional law also prohibits the creation of new types of taxes that do not align with those listed in Articles 105 and 106 GG.[30] A levy is considered a tax if it contributes to the general budget without functioning as payment for a state service, meaning it is payable unconditionally.[31]

As previously mentioned, the BEHG envisages an introductory phase from 2021 to 2026, during which fixed prices for emissions certificate trading are set. Under §5 (1) BEHG, the emission cap determined by §4 (1) BEHG may be exceeded during this introductory phase while adhering to a price corridor. Consequently, there is no shortfall in the cap in 2026 relative to the demand for emission allowances during the introduction phase and within the price corridor, as entities can purchase as many emissions certificates as they require. Thus, during this period, the national emissions trading system functions similiar to a tax and must be assessed in light of the existing financial constitutional requirements..[32]

Firstly, it is questionable whether CO_2 pricing should be construed as a tax or a non-tax levy. It represents an obligation to make a payment with a public character.[33] According to the legislators, it is a non-tax levy designed to capture the benefits of companies entitled to emit, grounded in the federal government's legislative competence on the subject.[34] However, the BVerfG has stated that the classification of a levy in terms of financial constitutionality does not depend on the legislator's designation but rather on its substantive content.[35]

---

[27] Klinski/Keimeyer 2020, p. 345.

[28] BVerfGE 93, 319 (342), BVerfGE 91, 186 (201); BVerfGE 82, 159 (178); Klinski/ Keimeyer, 2019, p. 1466; Rodi 2016, p. 532;

[29] See more: Klinski/ Keimeyer 2019, p. 1466 f.

[30] BVerfGE 145, 171; Klinski/ Keimeyer 2019, p. 1467; Klinski/ Keimeyer 2019, p. 1467.

[31] BVerfGE 124, 235 (243),

[32] Klinsli/Keimeyer 2019, p. 7 ff.; Antoni et al. 2019, p. 12f.

[33] Cf. BVerfG (Kammer) NVwZ 2018, 972 (974); Wernsmann/Bering 2020, p. 498.

[34] Bundestag, Drucksache 19/14746, 21, 37; Wernsmann/Bering 2020, p. 498 f.

[35] Cf. BVerfGE 149, 222 (250); Wernsmann/Bering 2020, p. 499; Klinski/Keimeyer 2020, p. 346.

The BVerfG states in its decision on groundwater utilisation fees that the revenue from the sale of emission allowances are not taxes but public levies. These levies are imposed as a public burden without individual consideration, i.e., unconditionally, to cover the general financial needs of a public community. However, the proceeds from these levies are not raised unconditionally but as consideration for the emission allowances acquired.[36] Applying this definition for the classification of the ETS from the period from 2027 onwards, it becomes evident that this type of levy can be regarded as a benefit skimming levy. The fees paid are in consideration for an individual benefit granted by the state—emission of CO_2 allowance.[37]

For the introductory phase until 2027, the situation should be evaluated differently. Benefit levies are justified as they compensate for a particular individual advantage. Without a distinct special advantage, which is the subject of the payment, the monetary payment becomes unconditional in the sense of a shared burden and, thus, should be classified as a tax.[38]

The BVerfG itself, in its ruling regarding the national emission trading, notes that the operator of an installation to whom the state sells emission allowances is granted a particular advantage over those operators of installations subject to emissions trading who do not possess sufficient emission allowances and are thus not allowed to use the air to the same extent for discharging CO_2 emissions. However, a sovereign granting of limited use rights is not a prerequisite for assuming a particular advantage.[39]

Clean air is a limited natural resource, capable of absorbing carbon dioxide only to a certain extent without harming the climate. Therefore, the limited good is not the air itself but its degree of pollution.[40] Consequently, even during the introduction phase, the legislator has designed the ETS based on the scarcity of available environmental resources determined by the state. Thus, the determination of a limited emission quota is predominant. The revenue from emission trading reflects the scarcity of these resources rather than their cause.[41]

The introduction period of the BEHG, without a fixed upper limit for the total quantity of emission allowances, implies that a fixed price system would be incompatible. If the total quantity of available emissions allowances is limited, then the allowance confers a particular benefit on its purchaser. Anyone receiving one of these limited allowances is granted a particular advantage, for which the state may legitimately demand consideration. In this scenario, the purchaser of the allowance does not obtain a specific

---

[36] BVerfG NVwZ 2018, 972 (974); Klinski/Keimeyer 2020, p. 346.

[37] Klinski/Keimeyer 2020, p. 346.

[38] Ibid, p. 347.

[39] BVerfG NVwZ 2018, 972 (974 f.); Klinski/Keimeyer 2020, p. 346.

[40] BVerfG NVwZ 2018, 972 (975).

[41] BVerfG NVwZ 2018, 972 (975); Klinski/Keimeyer 2020, p. 347.

individual advantage typical of a benefit levy but instead incurs a tax burden payable without preconditions.[42]

However, in academic literature, parallels are drawn to the BVerfg's "*Wasserpfenig*" decision[43] of the BVerfG are discussed to justify the benefit levy during the BEHG's introductory phase.[44] The distinctive feature of the benefit skimming levy is that it does not require state costs, but instead, a sufficient distinction from a tax is made through the dependence on consideration in the form of an individually allocable service. Companies purchasing emission certificates also receive an attributable advantage from the state compared to other companies that do not purchase such certificates and are thus not entitled to emit one tonne of greenhouse gas carbon dioxide equivalent in a certain period (as outlined in §3 (1) Nr. 2 BEHG). Therefore, an individual special advantage exists even during the introductory phase of the BEHG.[45]

Wernsmann and *Bering* criticise this point of view, arguing that the opportunity to use a natural environmental resource does not incur any costs for the state, even if the state permits individuals to manage environmental resources in a particular manner and then imposes a levy for it. Environmental resources such as air, water, and the atmosphere are public goods.[46] The public authorities cannot transform a cost-free good into a chargeable one for the citizens simply by imposing a fee, as this lacks the essential characteristic of state cost expenditure.[47] According to their perspective, the use of air and atmosphere should not be subject to CO_2 pricing under the BEHG, as it does not entail eligible state costs.[48]

The legal scholars *Jörn Ipsen, Nils Ipsen, Thomas Wisniewskin* and *Liesa Plappert* have expressed dpubts in their legal opinion[49] about whether the introductory phase of the BEHG can undergo an independent constitutional assessment. They argue that the introductory phase has a strong functional connection with the establishment of the national emissions trading system with market-based pricing, which is constitutionally conforming. Due to this solid functional link, the introductory phase cannot be viewed in isolation from the auction phase. The purpose of this phase is to facilitate a gradual transition from free use to market-price use, thus avoiding a price shock that would occur if a market price were introduced immediately.[50] This introductory phase also adheres to the principle of legal certainty, which requires, among other things, the predictability of state

[42] Klinski/Keimeyer, 2020 p. 347.

[43] BVerfGE 93, 319.

[44] Klinski/Keimeyer 2020, p. 347; Tappe 2020, p. 7;

[45] Tappe 2020, p. 8.

[46] Wernsmann/Bering 2020, p. 501.

[47] Cf. BVerfGE 144, 369 (397); Wernsmann/Bering 2020, p. 501.

[48] Wernsmann/Bering 2020, p. 501.

[49] Ipsen et al. 2022.

[50] Ibid, p. 12.

measures and actions, as well as the formal and intellectual accessibility of the law.[51] Nevertheless, this dispute can be left undecided at this point, as it is primarily theoretical, given that extending the BEHG to the building sector would necessitate amendments. The current legal concerns could potentially be addressed through these amendments.

However, doubts about legal uncertainty and the constitutionality of CO_2 pricing arise concerning the extendable price corridor set in §10 (2) BEHG. According to §23 (1) s. 4 f. BEHG, the Federal Government expressly reserves the right to decide on this in 2025. This means that if the Federal Government exercises this option, additional demand for emission allowances beyond the initially set cap could also be possible after the completion of the phase-in period. As long as a price corridor or cap exists, there is no binding cap, effectively extending the introduction phase.[52]

Moreover, the constitutional essentiality requirement (*Wesentlichkeitsvorbehalt*) might be violated, as the BEHG contains twelve ordinance authorisations. According to the BVerfG's case law derived from the rule of law principle (Art. 20 (3) GG) and the democracy principle (Art. 20 (1) and (2) GG), the legislature is obliged to make all fundamental normative decisions itself.[53] This principle is concretised by the requirement of certainty under Art. 80 (1) s. 2 GG, stipulating that the content, purpose, and extent of a regulatory authorisation must be determined in the law. The required degree of certainty depends on the nature of the facts to be regulated. In cases involving complex facts and foreseeable changes in circumstances, lower requirements for certainty are applicable.[54] Many of the BEHG's authorisations relate to complex issues with foreseeable changes, such as the calculation of annual emission quantities, or to the implementation of the BEHG, like the auctioning and trading of allowances. Hence, lower requirements for regulatory density apply here. Additionally, the enabling provisions predominantly contain specifics that limit the federal government's scope of action and offer some predictability. However, §11 (3) of the BEHG, with is open formulation and undefined legal terms, authorises the Federal Government to "regulate the measures necessary [...] to maintain the [...] competitiveness of affected companies". This provision does not allow individuals to define the extent of a regulatory authorisation established in the law. Nevertheless, as this legal provision concerns favourable administrative action, the requirements for certainty are also reduced. Therefore, the constitutional risks are minimal.[55]

Apart from the constitutional risks, the application of this financing instrument leaves the question open of how the financial resources would have to be distributed from the

[51] See Calmes-Brunet 2014, p. 602.

[52] Antoni et al. 2019, p. 13.

[53] BVerfGE 49, 89; Kotzur in: Münch et al. 2021, Art. 20 GG Rn. 1

[54] BVerfG, Urt. v. 19.09.2018, 2 BvF 1/15 a.a. O., Rn. 204; Antoni et al. 2019, p. 17.

[55] Antoni et al. 2019, p. 17.

federal government[56] to the landlords willing to undertake energy modernisation measures in the rented properties.

Building Renovation Tax

Hallof, in her dissertation, has developed a potential solution to the landlord-tenant dilemma in Germany,[57] with a focus on socially disadvantaged individuals. She suggests the introduction of a building renovation tax (or levy) applicable to anyone living and heating in Germany. This levy is conceptualised as a contribution to the use of the limited heat energy resource, representing a public transfer from citizens to the state. *Hallof* does not explicitly categorise this levy from a constitutional perspective. However, it is inferred that the Federation's competence to implement such a levy could be derived from its general competence under Articles 73 et seq. of the GG. *Hallof* terms this levy a tax, but in essence, it appears to be a special levy, as the competence for its implementation is suggested to be derived not from the Fiscal Constitution (Articles 104a et seq. GG) but from the general provisions regarding the distribution of subject-matter competencies. This proposed levy would serve as a specific charge to fund energy building renovations and would be imposed on both tenants and owners occupying their premises. An alternative approach could involve charging the levy to private households as a whole, rather than to individuals.[58]

Currently, there is no special levy on CO_2 emissions in Germany.[59] Introducing CO_2 pricing and emission trading would be generally possible in Germany. When formulating a specific legal framework for such an instrument, it is crucial to ensure that any interactions with existing systems, such as the EU ETS, are carefully considered. Failure to do so could lead to unintended consequences that might undermine the effectiveness of the new CO_2 pricing mechanism.[60]

According to the solution proposed by *Hallof*, the levy should be uniform for all individuals. There are several ways to structure this: it could be a per capita charge, or the charge could be linked to the taxpayer's age. Another option is to link the charge to the size of the occupied apartments, using the unit value of the living space as the basis for taxation. To support social equity, recipients of social benefits, students, and similar groups could receive subsidies funded through general taxation.[61]

The primary objective of the building renovation levy is to facilitate the renovation of the building stock, not to generate revenue for the government. Once the building stock is comprehensively renovated, the levy should be discontinued. Exemptions or tax

---

[56] ESRa P4.3., p. 24.

[57] Hallof 2013, p. 204f.

[58] Ibid, p. 204.

[59] Antoni/Rodi 2019, p. 16.

[60] Antoni/Rodi 2019, p. 19; see more: Antoni/Rodi 2019, p. 6f.

[61] Hallof 2013, p. 205.

reductions should be considered for individuals already bearing the costs of energy-related renovations and for those whose energy consumption falls below certain thresholds. Alternatively, the building renovation charge could be time-limited or tied to the achievement of specific energy efficiency targets.[62]

Hallof emphasises the importance of providing subsidy entitlements to those who shoulder the economic burden of energy renovation. The revenue generated from the charge could support a funding programme for efficient energy renovation, benefiting both landlords and owner-occupiers. The subsidy should not cover the entire cost of construction measures, but rather a reasonable base amount determined based on the classification of the building. Factors for this classification could include the building's age, its thermal conditions, regional and structural characteristics of the location, and its rentability (for instance, vacant buildings in areas of emigration may not require renovation). Additionally, incentives should be created for landlords to keep construction costs low and opt for the most efficient renovation measures.[63]

The effectiveness of building renovation lies in creating incentives for all stakeholders in the housing market. Tenants would not have to worry about significant rent increases following housing renovations. With a uniform renovation tax for all tenants, there would be no financial benefit in choosing an unrenovated apartment. Consequently, this would increase the pressure on landlords to renovate, as tenants would prefer living in renovated apartments with lower heating costs if they are already paying a building renovation charge.[64]

However, as previously mentioned, *Hallof*'s proposed solution could be classified as a special levy, which poses constitutional challenges. Special levies may conflict with the constitutional principle of the "tax state", which stipulates that the state should primarily meet its financial needs through taxes.[65] The BVerfG recognises special levies only as a rare exception,[66] as they could undermine the federal distribution of competencies under fiscal constitutional law.[67] Articles 104a et seq. of the GG establish a differentiated distribution of legislative, revenue, and administrative competencies for taxation. The Federal Constitution does not explicitly address the admissibility of other non-tax forms of levies or special levies. Therefore, it should be assumed that the GG does not contain a definitive list of permissible types of levies, implying that non-tax forms of levies must be based on the general provision of subject matter competence under Articles 70 et seq. GG. However, if both the Federation and the *Länder* were permitted to levy any type of tax in addition to those under the Fiscal Constitution (Articles 104a et seq. GG), the

[62] Ibid.

[63] Ibid, p. 206f.

[64] Ibid, p. 209f.

[65] BVerfGE 93, 319 (342); ESRa P4.3., p. 20; Elsner/Kaltenborn 2005, p. 824.

[66] BVerfGE 55, 274 (308); BVerfGE 82, 159 (181).

[67] BVerfG, 03.02.2009 – 2 BvL 54/06; ESRa P4.3., p. 20.

distribution of competencies established therein would largely lose its significance and function.[68]

According to the case law of the Federal Constitutional Court, a specific objective justification is necessary for special levies to avoid conflict with the principle of fiscal statehood.[69] Moreover, a special levy imposes a particular financial burden on specific groups of citizens, thereby creating an inequality of burdens. While this thesis does not examine the compatibility of such levies with fundamental rights, it is noted that a special levy may only be imposed on social groups that are "separable from the general public and other groups by a common interest situation predetermined by the legal system or social reality or by special common circumstances".[70] This targeted social group must be homogeneous, such as members of a specific professional group. Otherwise, the legislator would be granted excessive latitude in determining the groups especially burdened by the levy, risking the replacement of general taxes with special levies.[71]

The target group for the levy proposed by *Hallof*, which includes everyone living in Germany and using heating, presents challenges regarding this homogeneity criterion. While the legislator has some discretion in defining homogeneous groups for special levies, the group proposed by *Hallof* is excessively broad and lacks a specific common interest that separates it from the general public or other groups. It cannot be assumed that the mere use of heating systems constitutes a common interest that unites this group. Households in Germany that use heating do not necessarily perceive themselves as connected, nor have they implicitly agreed to distinguish themselves from other social groups. Therefore, *Hallof*'s proposed group for the special levy may not meet the necessary criteria for homogeneity and separability from the general public as required by constitutional principles.

Energy Tax Reform

Another potential solution to the landlord-tenant dilemma is the transformation of the current energy tax into a combined CO_2 energy tax.[72] However, it is important to note that this instrument would have an exclusive relationship with the national emission trading system, as both target the same taxable event: CO_2 emissions. In Germany, the consumption of fossil fuels is currently taxed under the Energy Tax Act (EnergieStG) and the Electricity Tax Act (StromStG),[73] with the energy tax functioning as a consumer tax (§1 (1) s. 3 EnergieStG). Presently, only energy sources fall under the Energy Tax Act,

---

[68] BVerfGE 55, 274 (300f.); BVerfGE 108, 186 (215); Elsner/Kaltenborn 2005, p. 824.

[69] BVerfG, 03.02.2009 – 2 BvL 54/06; ESRa P4.3., p. 20.

[70] BVerfGE 55, 274 (305 f.); Elsner/Kaltenborn 2005, p. 827.

[71] Elsner/Kaltenborn 2005, p. 827.

[72] Rodi et al. 2016, p. 187ff; Rodi 2018; Steinbach/Valta 2019, p. 9f.

[73] Antoni/Rodi 2019, p. 6.

encompassing all fuels and heating materials.[74] An additional tax on CO_2 emissions could significantly enhance the ecological steering effect.

Attempts have been made at the EU level to introduce a uniform Europe-wide CO_2 tax, but these have not been successful to date.[75] Some EU member states, like Sweden, have been taxing CO_2 emissions nationally for many years. Sweden, with over two decades of experience in CO_2 taxation, had the world's highest CO_2 tax in 2010. Concurrently, Sweden's climate protection targets for 2020 were among the most ambitious globally. However, Sweden also maintains an energy tax, resulting in both fossil CO_2 emissions and energy being taxed.[76]

For Germany, a combined CO_2 energy tax would be more constitutionally viable. The BVerfG denied in its decision to the nuclear fuel tax (*Beschluss zur Kernbrennsteuer*)[77] the competence of the Federal legislature to invent taxes that has been disputed for decades.[78] The legal issue was whether the federal government could derive a competence to introduce the nuclear fuel tax from Article 105 (2) GG, i.e., whether this tax is an "other tax" as per this provision. The BVerfG concluded that "other taxes" refer only to those mentioned in Article 106 GG, implying that the Federal Constitution has only a *numerus clausus* of taxes and their types, providing a framework for the legislator in inventing new taxes.[79]

For taxes and types of taxes listed in Art. 105 and 106 GG, specific typological concepts are applied. The respective average case is considered to determine the characteristics defining a tax type, excluding features unique to individual phenomena. The constitutional interpretation of these articles does not require that all characteristics defining the type are always present. When introducing new tax concepts, classification depends on the principles of traditional German tax law. Thus, in developing a new tax that conforms constitutionally, it is necessary to identify those characteristics typically associated with a tax or type of tax and necessary for distinguishing it from other taxes or types of taxes.[80] The legislature has no discretion in regulating the tax allocation it introduces, a limitation protecting citizens from being overburdened by the unrestricted introduction of new taxes.[81]

A pure CO_2 energy tax would be difficult to classify as a consumer tax from a constitutional point of view. The BVerfG has developed the following features for

---

[74] Stein et al. 2016, p. 47.

[75] 'Verbrauchsteuern: Kommissionsvorschlag zur Energiebesteuerung - Europäische Kommission'. n.d. Accessed 24 December 2023. https://taxation-customs.ec.europa.eu/excise-duties-energy-tax-proposal_de. ; Rodi 2018, p. 744.

[76] Hammar/Sjöström 2011, p. 6672f.

[77] BVerfGE 145, 171.

[78] Leisner-Egensperger 2019, p. 2220.

[79] Leisner-Egensperger 2019, p. 2220 f.; BVerfGE 145, 171 (193f.).

[80] BVerfGE 145, 171 (193).

[81] Leisner-Egensperger 2019, p. 2221.

distinguishing consumer taxes among other taxes and types. The GG does not define a consumer tax; it is broadly understood as a tax type. They are to be distinguished from corporate taxes that are not based on the use of income but on the generation thereof. A consumer tax is an indirect tax levied on the producer and designed to be passed on to the (end) consumer. Moreover, the consumption of durable goods is required for this type. Furthermore, consumer taxes are regularly linked to the transfer of consumer goods from a tax nexus to general economic transactions, which are not tax-bound.[82] According to these features, the pure CO_2 tax cannot be subsumed as a consumer tax, and it could consequently be unconstitutional. Given the lack of a tax invention right, the nuclear fuel tax was already declared unconstitutional by the Federal Constitutional Court for this reason. However, a combined CO_2 energy tax would remain a consumer tax and thus be permissible under Art. 105 (2) GG in conjunction with Art. 106 GG.[83]

Reforming the existing energy tax into a two-component tax, which includes both taxation of energy and CO_2 emissions, would help overcome the constitutional limitations described above. Such a reform would achieve a decarbonisation steering effect within the framework of an existing consumer tax. The energy tax is a classic consumer tax, which is indirect, meaning the tax debtor is not necessarily the same as the taxpayer. The tax is usually passed on by the tax debtor to another person, for example, contractually. The reform of the energy tax to a combined tax would link the tax to the CO_2 intensity of an energy source, such as coal.[84] The CO_2 component within an energy consumption tax framework would directly address climate protection, especially in the heat and transport sectors, where existing instruments have proven deficient,[85] as well as in the building sector. One possibility is to set the level of energy tax rates according to the consumption of an energy source, at least proportionally to the associated CO_2 emissions—this would be the CO_2 component.[86]

Reforming the energy tax into a combined CO_2 energy tax would serve the dual objectives of resource protection policy, both economic and ecological. The state could reinvest the additional costs in energy-efficient building refurbishment, for example, through interest-free loans or subsidies. These pecuniary benefits could be granted to citizens by the public administration authority for modernisation measures, thereby promoting the achievement of climate protection and energy efficiency goals.[87]

The CO_2 component could enhance energy efficiency in two ways. Firstly, taxing CO_2 emissions would encourage more environmentally friendly behavior by polluters. This is

---

[82] BVerfGE 145, 171 (211).

[83] BVerfGE 145, 171, with dissenting opinion of Judges Huber and Müller in a separate opinion; Rodi 2018, p. 748; Franzius 2018, p. 1591.

[84] Antoni/ Rodi 2019, p. 6.

[85] Rodi 2018, p. 747 f.

[86] Antoni/ Rodi 2019, p. 6.

[87] Rodi 2014, p. 217; Antoni/ Rodi 2019, p. 6.

because the correlation between fossil fuel use and fossil CO_2 emissions for the respective fuel becomes more apparent.[88] Secondly, emission reductions could be achieved through extensive energy-efficient modernisation measures in the residential building sector. Furthermore, the reformed energy tax could internalise the external effects of rental contract parties by shifting the financing of modernisation measures to the state. This measure is also essential considering the risks of climate change for future generations.[89] The proposed scheme would effectively break the vicious circle of the landlord-tenant dilemma.

In the medium-term, an improvement of the housing stock in the rental housing market could be expected. This would likely reduce competition between residential properties, as rents in renovated buildings are often higher than in those that have not been renovated. Moreover, maintenance of rents at a lower level is more likely, creating a rent control effect. Socially disadvantaged households, who often live in energy-inefficient buildings,[90] could also afford a more energy-efficient flat. In this way, the state's social obligations would also be addressed.

However, reforming the existing tax is time-consuming, as a law must pass through all levels of parliament and requires majority support.[91] Moreover, considering the exclusive relation between the national emission trading system, which prices CO_2 emissions, and the energy tax reform discussed above, the BEHG would need to be repealed. Subsequent gaps in the legislation would have to be filled by further legislative reforms.

Concept of Partial Warm Rent as Reform of Heating Cost Allocation

Considering the current socio-political situation, researchers from various fields are focusing on developing incentives for energy-efficient building refurbishment to meet Germany's energy efficiency targets and reduce CO_2 emissions. Environmental economics and environmental law offer a range of regulatory instruments to internalise external effects and address the landlord-tenant dilemma. Literature in these fields suggests introducing the concept of partial warm rent.[92]

The allocation of utility costs, especially for heating and electricity, is important in examining the cost distribution between tenants and landlords for energy modernisation measures, in addition to general tenancy law regulations. The definition of utility costs is established in §556 (1) s. 2 BGB. According to this, utility costs are ongoing expenses incurred by the owner or leaseholder due to ownership, the heritable building right to the

[88] Hammar/ Sjöström 2011, p. 6675.

[89] Eidenmüller 2015, p. 155.

[90] Boardman 2010, p. 47f.

[91] ESRa P4.3., p. 20.

[92] Knissel et. al., 2001 p. 35f.; Freund 2018, p. 86 et seq.; Klinski/ Keimeyer/ Braungardt 2021.

land, or the intended use of the building, outbuildings, installations, facilities, and the land.[93]

The transition to a partial warm rent model primarily targets existing buildings. Its objective is to incentivise landlords within the tenancy relationship to reduce heating energy consumption, thereby improving the energy quality of the building. Such an incentive is currently absent in the cold rent system, where landlords pass on higher energy costs to tenants as separate items. As a result, landlords do not directly experience the impact of rising energy costs.[94]

Different models of partial warm rent are being explored in literature and research, all sharing the common goal of splitting costs between the landlord and tenant to incentivize both parties to contribute to energy efficiency. The models propose dividing heating costs into user-dependent and basic (fixed) heating costs. A building with high heating costs would have a high partial warm rent, while a building with low heating costs would have a low partial warm rent.[95] The fixed heating costs are then transferred to the landlord, who is responsible for providing dwellings and the ongoing heat supply. The difference in these models lie in the criteria used to determine the fixed heating costs and how they are measured.

One such model is the *reference temperature model*, used also in some housing estates in Sweden, though it is not legally binding there. This model guarantees a minimum temperature covered by the fixed warm rent.[96] The basic heating costs are combined with the cold rent to form the partial warm rent. Under this system, the landlord would be responsible for providing basic heating, for example, up to 20 degrees. Additional energy consumption for heating the flat is charged according to usage, introducing the "polluter pays" principle. The tenant bears the costs arising from individual heating behavior, while the landlord covers the costs related to the energy condition of the building.[97] This approach, proposed for adoption in Germany by Agora Energiewende, is known as "*Warmmieten mit Temperaturfeedback*" ("warm rent with temperature feedback"),[98] ensuring that tenants do not experience increased financial burdens from energy refurbishment, thus preserving warm rent neutrality.[99]

Another alternative, introduced by *Klinski, Keimeyer*, and *Braungardt*, is the "*Prozentmodell*" ("percentage model"), which produces similar effects to the reference temperature model. The key difference is that the regulatory components are based on the consumption costs of heating energy, split between the landlord and tenant at a

[93] BBSR 2016, p. 95.

[94] Klinski/ Keimeyer/ Braungardt 2021, p. 3.

[95] Freund, p. 87 f.; Klinski/ Keimeyer/ Braungardt 2021, p. 5.

[96] Klinski/ Keimeyer/ Braungardt 2021, p. 5 f.

[97] Freund 2018, p. 87 f.

[98] Agora Energiewende und Universität Kassel 2020, p. 12 f.

[99] Freund 2018, p. 87 f.

predetermined percentage rate (e.g., 50:50 or 70:30). This split forms a fixed component for the future partial warm rent, and the remaining percentage share of the consumption costs is allocated to the tenant.[100]

Under the current legal regime of the cold rent, tenants are required to bear the costs of CO_2 pricing, which is anticipated to increase in the coming years. However, the consumption of heating energy primarily depends on the structural and technical condition of the building, a domain that falls under the landlord's responsibility and is beyond the tenant's influence. Shifting these costs to the landlord would align with the allocation of liability in tenancy law. According to §535 (1) BGB, landlords must ensure that the rented property is in a condition suitable for use as stipulated in the tenancy agreement. The amount of greenhouse gas emissions is largely determined by the structural and technical attributes of the dwelling and building, for which landlords hold responsibility. This is because they are tasked with maintaining the dwelling in a condition that is suitable for everyday use.[101]

6.2.2 Sweden

6.2.2.1 Green Bonds

Swedish municipalities are actively engaged in environmental and social sustainability efforts, as well as striving towards climate neutrality. However, there is no national coordination in this regard, leading to significant variation in ambitions, measures, and approaches across different municipalities.[102] Despite this, some practices being implemented are of interest not only to Sweden but also to other countries. A notable example is the use of municipal and corporative green bonds as instruments for debt financing.

Green bonds function similarly to conventional bonds in that they both serve to mobilise funds for specific projects or ongoing needs of the issuer. The issuer commits to making fixed, regular payments and fully reimbursing the bond at maturity. The key distinction is that green bonds carry a 'green' label, intended to assure investors that the funds raised will be used for environmentally and climate-relevant projects. However, since issuers can self-label their bonds as 'green', this designation is not always reliable and can sometimes be used for 'greenwashing'. Recognizing this issue, financial market actors have developed numerous standards for green bonds at the international level, such as the Green Bond Principles and the Climate Bond Standards.[103] Additionally, a

[100] Klinski/ Keimeyer/ Braungardt 2021, p. 6.

[101] Klinski/ Keimeyer/ Braungardt 2021, p. 3 f.; Agora Energiewende and Universität Kassel 2020, p. 11 f.

[102] Sjöström et al. 2020, p. 6.

[103] Karras, ZKF 3/2022, p. 55; IRENA 2020, p. 6.

taxonomy has been adopted at the EU level to provide further guidance and standardisation in this area (see Sect. 4.3.3).

In general, the role of municipalities is crucial in achieving the Sustainable Development Goals, as 85% of the Swedish population resides in cities and urban areas. Accessing finance poses a significant challenge for cities implementing climate change plans. Several Swedish municipalities have issued green bonds to mobilise private investment into sustainable projects.[104] The green bond market has expanded over the past decade, with Sweden being an early adopter in this arena.[105] Among the Nordic countries, Sweden has been having the largest corporate bond market and the most extensive local bond market since 2014.[106] The city of Gothenburg was a pioneer in issuing a municipal green bond in 2013, funding low-carbon and climate-resilient transition projects. Subsequently, Stockholm Läns Landsting, Skåne Region, and the cities of Malmö, Lund, Västerås, Örebro, along with other Swedish municipalities, have issued municipal green bonds as well.[107]

As previously mentioned, municipalities bear responsibility for housing provision in Sweden. In 1986, nine Swedish municipalities and Örebro County Council established Kommuninvest, a Swedish Local Government Funding Agency, to enhance conditions in the local government loan financing market. In 2020, 60% of Kommuninvest's green portfolio was allocated to green buildings.[108] As of 31 December 2021, Kommuninvest financed 292 green building projects through green bonds, including five energy efficiency projects in existing buildings and 25 major modernisations of existing structures.[109] Given the significant proportion of municipally-owned dwellings in Sweden's residential tenancy market, covering 20% of all dwellings,[110] this approach could markedly influence the further decarbonisation of the residential building sector.

However, Sweden is also famous for mobilising private funds for green debt financing in the corporative sector. Sweden's largest real estate company Vasakronan possesses properties in Stockholm, Upsala, Gothenburg and Malmö[111] has been the world's first company to issue a (corporative) green bond in November 2013.[112] The company

---

[104] Vanhuyse/Chan 2020, p. 3 f.

[105] Torvanger/ Maltais/ Marginean 2021, p. 1.

[106] Climate Bond Initiative, 2018, p. 6.

[107] Vanhuyse/Chan 2020, p. 4.

[108] h Cicero 2021, p.5.

[109] Kommuninvest Green Bonds Impact Report 2021, p. 24.

[110] 'Just over 5 Million Dwellings in Sweden'. n.d. Statistiska Centralbyrån. Accessed 24 December 2023. https://www.scb.se/en/finding-statistics/statistics-by-subject-area/housing-construction-and-building/housing-construction-and-conversion/dwelling-stock/pong/statistical-news/dwelling-stock-2020-12-31/.

[111] Karras/Novikova 2021, p. 29.

[112] Vasakronan 2018a, b, p. 2.

developed a sustainability programme with specific environmental, financial, and social performance targets and reporting via the Global Reporting Initiative.[113] This enabled investors possibility to track portfolio development and ensure targeted insight.[114] Vasakronan included environmental requirements in all the agreements with its 2,800 direct suppliers recognising environmental and social consequences across the life cycle of building design, construction, management, and demolition, including the supply chain.[115] Building energy retrofit upgrade to a more efficient one is one of the eligible assets of the Green Asset Pool of the company.[116] As a result of these initiatives, carbon emissions from Vasakronan's portfolio were reduced by 97% between 2006 and 2013, with energy consumption decreasing by 30% in the same period.[117]

Green bonds serve as an effective tool for economically challenged regions, addressing the landlord-tenant dilemma since financing originates from a third party, not the landlord or tenant. This mechanism enables the raising of funds for the energy modernisation of dwellings by municipalities and corporate firms owning rental properties. Such an additional financial source for energy modernisation is particularly important for socially vulnerable groups with limited means to contribute financially to their dwellings.[118]

References

Antoni, Johannes, and Michael Rodi. 2019. 'Transformation Des Stromsystems. Möglichkeiten Einer Flankierenden CO_2-Bepreisung Durch Öffentlich-Rechtliche Abgaben'. https://www.ikem.de/wp-content/uploads/2019/02/20190129_CO2-Bepreisung_Kohleausstieg_ENavi_Paper.pdf.

Boardman, Brenda. 2010. *Fixing Fuel Poverty: Challenges and Solutions*. London. https://www.routledge.com/Fixing-Fuel-Poverty-Challenges-and-Solutions/Boardman/p/book/9781844077441.

BVerfG, 03.02.2009 – 2 BvL 54/06 – CMA-Pflichtabgabe, Absatzfonds. 2009. Urt. BVerfG.

BVerfG, 5.3.2018 – 1 BvR 2864/13 – Kürzung der Zuteilung kostenloser Emissionsberechtigungen verfassungsgemäß. 2018. Beschl., 2018 NVwZ 972–78. BVerfG.

BVerfG, 07.07.2009 – 1 BvR 1164/07 – Betriebsrente, Homo-Ehe, Hinterbliebenenversorgung. 2009. Beschl. BVerfG.

BVerfG, 07.11.1995 – 2 BvR 413/88, 2 BvR 1300/93 – Wasserpfennig. 1995. Beschl. BVerfG.

BVerfG, 08.08.1978 – 2 BvL 8/77 – Kalkar I, Schneller Brüter. 1978. Beschl. BVerfG.

[113] Karras/Novikova 2021, p. 29.

[114] CICERO 2018, p. 8.

[115] Ibid, p. 6.

[116] Vasakronan 2018a, b, p. 4.

[117] Cicero, p. 6; Karras/Novikova 2021, p. 29.

[118] Karras/Novikova 2021, p. 28f.

BVerfG, 10.12.1980 – 2 BvF 3/77 – Berufsausbildungsabgabe, Ausbildungsplatzförderung. 1980. Urt. BVerfG.

BVerfG, 11.10.1994 – 2 BvR 633/86 – Kohlepfennig. 1994. Beschl. BVerfG.

BVerfG, 13.04.2017 – 2 BvL 6/13 – Kernbrennstoffsteuer. 2017. Beschl. BVerfG.

BVerfG, 17.01.2017 – 2 BvL 2/14, 2 BvL 3/14, 2 BvL 4/14, 2 BvL 5/14 – Rückmeldegebür Brandenburgisches Hochschulgesetz. 2017. Beschl. BVerfG.

BVerfG, 31.05.1990 – 2 BvL 12/88, 2 BvL 13/88, 2 BvR 1436/87 – Absatzfonds. 1990. Beschl. BVerfG.

Calmes-Brunet, Sylvia. 2014. 'Calmes-Brunet: Rechtssicherheit Und Vertrauensschutz Im Verfassungsrecht'. *Juristische Schulung*, no. 7: 602–6.

CICERO. 2018. '"Second Opinion" on Vasakronan's Green Finance Framework'. https://pub.cicero.oslo.no/cicero-xmlui/bitstream/handle/11250/2720386/vasakronan_2018.pdf?sequence=1&isAllowed=y.

Claudio. 2018. 'Rechtsprobleme Des Kohleausstiegs'. *Neue Zeitschrift Für Verwaltungsrecht*, no. 21: 1585–91.

Climate Bond Initiative. 2018. 'The Green Bond Market in the Nordics 2018'. https://www.climatebonds.net/files/files/CBI-The%20Green%20Bond%20Market%20in%20the%20Nordics.pdf.

Communication from the Commission to the European Parliament, the European Council, the Council, the European Economic and Social Committee and the Committee of the Regions the European Green Deal. 2019. https://eur-lex.europa.eu/legal-content/EN/TXT/?uri=COM%3A2019%3A640%3AFIN.

Deutscher Bundestag. 2019. *Gesetzesentwurf Der Fraktionen CDU/CSU Und SPD, Entwurf Eines Gesetzes Über Einen Nationalen Zertifikatehandel Für Brennstoffemissionen (Brennstoffemissionshandelsgesetz – BEHG).*

Deutscher Bundestag. 2020. 'Antrag Der Abgeordneten Und Der Fraktion Der FDP, Fairer Klimaschutz Für Mieter Und Vermieter'. Drucksache 19/25246. https://dserver.bundestag.de/btd/19/252/1925246.pdf.

Eidenmüller, Horst. 2015. *Effizienz Als Rechtsprinzip*. 4th ed. Vol. 90. Die Einheit Der Gesellschaftswissenschaften. Tübingen: Mohr Siebeck.

Elsner, Bernd R., and Markus Kaltenborn. 2005. 'Sonderabgaben Im Steuerstaat'. Juristische Arbeitsblätter, no. 11: 823–28.

Entwurf Eines Gesetzes Zur Änderung Des Gesetzes Über Die Kreditanstalt Für Wiederaufbau Und Weiterer Gesetze, Gesetzesentwurf Der Bundesregierung. 2013. https://dserver.bundestag.de/brd/2013/0215-13.pdf.

'ESRa – Global Climate Forum'. n.d. Accessed 24 November 2023. https://globalclimateforum.org/portfolio-item/esra/.

'Förderprodukte für energieeffiziente Sanierung – Übersicht | KfW'. n.d. Accessed 24 December 2023. https://www.kfw.de/inlandsfoerderung/Privatpersonen/Bestandsimmobilie/Energieeffizient-Sanieren/Förderprodukte/.

Franzius, Claudio. 2022. 'Der „Green Deal" in Der Mehrebenenordnung'. *Klima Und Recht*, no. 1: 2–5.

Freund, Stefan. 2018. *Die Berücksichtigung der energetischen Ausstattung und Beschaffenheit in qualifizierten Mietspiegeln: Bestandsaufnahme, empirische Analyse und Empfehlungen.* 1st ed. Siegburg Köln: Josef Eul Verlag.

'Gesetz Über Einen Nationalen Zertifikatehandel Für Brennstoffemissionen (Brennstoffemissionshandelsgesetz – BEHG)'. 2019. *Bundesgesetzblatt Teil I*, no. 50 (December): 2728.

Hammar, Henrik, and Magnus Sjöström. 2011. 'Accounting for Behavioral Effects of Increases in the Carbon Dioxide (CO_2) Tax in Revenue Estimation in Sweden'. *Energy Policy*, Sustainability of biofuels, 39 (10): 6672–76. https://doi.org/10.1016/j.enpol.2011.06.014.

Ipsen, Nils, Thomas Wisniewski, Liesa Plappert, and Jörn Ipsen. 2022. 'Zur Verfassungsmäßigkeit Des Brennstoffemissionshandelsgesetzes (BEHG)'. Rechtsgutachten. Dessau-Roßlau: UBA. https://www.umweltbundesamt.de/sites/default/files/medien/479/publikationen/texte_04-2022_zur_verfassungsmaessigkeit_des_brennstoffemissionshandelsgesetzes.pdf.

IRENA. 2020. 'Green Bonds'. Renewable Energy Finance Brief 03. https://www.irena.org/-/media/Files/IRENA/Agency/Publication/2020/Jan/IRENA_RE_finance_Green_bonds_2020.pdf.

Karras, Jana. 2022a. 'Kommunale Green Bonds Im Lichte Der Schuldenbremse – Teil 1'. *ZKF*, no. 3: 54–58.

Karras, Jana. 2022b. 'Kommunale Green Bonds Im Lichte Der Schuldenbremse – Teil 2'. *ZKF*, no. 4: 80–84.

Karras, Jana, and Aleksandra Novikova. 2021. 'Germany. Legal Barriers to the Decarbonisation of Dwellings Occupied by Low-Income Tenants and Financial Incentives to Overcome These Barriers'. IKEM Climate Finance Week. https://www.ikem.de/wp-content/uploads/2021/05/Brief_5_Legal-barriers-to-the-decarbonisation-of-dwellings.pdf.

Keimeyer, Friedhelm, and Stefan Klinski. 2020. 'Zur Finanzverfassungsrechtlichen Zulässigkeit Der CO_2-Bepreisung Nach Dem Brennstoffemissionshandelsgesetz (BEHG)'. *ZUR*, no. 6: 342–50.

KfW. 2020. 'Anlage Zu Den Merkblättern Energieeffizient Sanieren – Kredit Und Investitionszuschuss'. 151/152/430. https://www.kfw.de/PDF/Download-Center/Förderprogramme-(Inlandsförderung)/PDF-Dokumente/6000003612_M_151_152_430_Anlage_TMA_2018_04.PDF.

Klinski, Stefan, and Ass Friedhelm Keimeyer. 2019a. 'Die Besteuerung von CO2 Und Andere Umweltabgaben Im Lichte Des Finanzverfassungsrechts'. *Neue Zeitschrift Für Verwaltungsrecht*, no. 20: 1465–71.

Klinski, Stefan, and Ass Friedhelm Keimeyer. 2019b. 'Zur Finanzverfassungsrechtlichen Zulässigkeit Eines Nationalen Zertifikathandels Für CO2-Emissionen Aus Kraft- Und Heizstoffen, Rechtswissenschaftliches Kurzgutachten'. https://www.oeko.de/fileadmin/oekodoc/Verfassungsrecht_Emissionshandel_Gebaeude-Verkehr.pdf.

Knissel, Jens, Roland Alles, Iris Behr, Eberhard Hinz, Tobias Loga, and Joachim Kirchner. 2001. 'Mietrechtliche Möglichkeiten Zur Umsetzung von Energiesparmassnahmen Im Gebäudebestand'. Frankfurt am Main: IWU. https://www.dbu.de/OPAC/ab/DBU-Abschlussbericht-AZ-15609.pdf.

Kommuninvest. 2021. 'Green Bonds Impact Report'. https://kommuninvest.se/wp-content/uploads/2022/04/Green-Bonds-Impact-Report-2021.pdf.

Kreuter-Kirchhof, Charlotte. 2022. 'Kohärenz Der Klimaschutzinstrumente'. *Klima Und Recht*, no. 3: 70–74.

Leisner-Egensperger, Anna. 2019. 'CO2-Steuer Als Klimaschutzinstrument'. *Neue Juristische Wochenschrift*, no. 31: 2218–21.

Novikova, Tim Langenhorst, and Jana Karras. n.d. 'Systematisierter Und Kategorisierter Juristischer Instrumentenkanon, Energiewende Im Sozialen Raum'. Produkt No. 4.3. https://esra-projekt.de/sites/default/files/2022-02/2021.09.14_esra_0.pdf.

Rodi, Michael. 2014. *Ökonomische Analyse des Öffentlichen Rechts*. Lehrbuch. Berlin Heidelberg: Springer Gabler.

Rodi, Michael. 2016. 'Der Rechtsrahmen Für Den Einsatz Ökonomischer Instrumente in Der Ressourcenschutzpolitik'. *ZUR*, no. 10: 531–41.

Rodi, Michael. 2018. 'Die Zukunft Der Energiesteuern Im Rahmen Der Energiewende'. IKEM. https://www.ikem.de/wp-content/uploads/2018/08/GS_Joecks_Rodi.pdf.

Rodi, Michael, Erik Gawel, Alexandra Purkus, and Anne Seeger. 2016. 'Energiebesteuerung Und Die Förderziele Der Energiewende – Der Beitrag von Energie- Und Stromsteuern Zur Förderung von Erneuerbaren Energien, Energieeffizienz Und Klimaschutz'. *STuW*, 187–99.

Senatsverwaltung für Stadtentwicklung und Umwelt. n.d. 'Klimaneutrales Berlin 2050. Empfehlungen Für Ein Berliner Energie- Und Klimaschutzprogramm (BEK)'. Accessed 19 December 2021. https://www.berlin.de/sen/uvk/klimaschutz/klimaschutz-in-der-umsetzung/das-berliner-energie-und-klimaschutzprogramm-bek/der-weg-zum-bek/#bek-konsolidiert.

Sjöström, Emma, Nina Waltré, Astrid Nilsson, Alicia Requena Carrion, Aina Eriksson, Magnus Agerström, and Fedra Vanhuyse. 2020. 'Financing Sustainable City Goals in Sweden; Preliminary Findings'. Discussion brief. https://www.sei.org/wp-content/uploads/2020/10/discussion-brief-final-1.pdf.

Stein, Roland M., Anahita Thoms, and Sabine Schröer-Schallenberg. 2016. *Energiesteuern in der Praxis: Energiesteuer – Stromsteuer – Biokraftstoffquote – Energiesteuer-Compliance*. 3., Aktualisierte und überarbeitete Auflage. AW-Prax Schriftenreihe. Köln: Bundesanzeiger Verlag.

Steinbach, Armin, and Matthias Valta. 2019. 'CO2-Orientierte Bepreisung Der Energieträger – Handlungsoptionen, Kompensationsmöglichkeiten Und Ihre Rechtlichen Rahmenbedingungen'. DICE Ordnungspolitische Perspektiven 101. Heinrich Heine University Düsseldorf, Düsseldorf Institute for Competition Economics (DICE). https://econpapers.repec.org/paper/zbwdiceop/101.htm.

Tappe, Henning. 2020. 'PA 16/002 – 5410 – Öffentliche Anhörung Am 16. September 2020 Zu Dem Gesetzentwurf Der Bundesregierung Eines Ersten Gesetzes Zur Änderung Des Brennstoffemissionshandelsgesetzes'. Drsn. 19/19929, 19/21755. Universität Trier. https://www.bundestag.de/resource/blob/791380/94b0b71d634a94087323493dcc181b44/Universitaet-Trier-Prof-Dr-Tappe-data.pdf.

Torvanger, Asbjørn, Aaron Maltais, and Iulia Marginean. 2021. 'Green Bonds in Sweden and Norway: What Are the Success Factors?' *Journal of Cleaner Production* 324 (November): 129177. https://doi.org/10.1016/j.jclepro.2021.129177.

Vanhuyse, Fedra, and Stefanie Chan. 2020. 'Bonds beyond Green', March. https://www.sei.org/publications/bonds-beyond-green/.

Vasakronan. 2018a. 'Vasakronan Green Finance Framework'. Stockholm. https://wwwvasakronanse.cdn.triggerfish.cloud/uploads/2018/10/vasakronan_green_bond_framework.pdf.

Vasakronan. 2018b. 'VASAKRONAN GREEN FINANCE FRAMEWORK'. https://wwwvasakronanse.cdn.triggerfish.cloud/uploads/2018/10/vasakronan_green_bond_framework.pdf.

'Verbrauchsteuern: Kommissionsvorschlag zur Energiebesteuerung – Europäische Kommission'. n.d. Accessed 24 December 2023. https://taxation-customs.ec.europa.eu/excise-duties-energy-tax-proposal_de.

Wernsmann, Rainer. 2020. 'Verfassungswidrigkeit Der CO2-Bepreisung Nach Dem Brennstoffemissionshandelsgesetz (BEHG), Insbesondere Im Zeitraum von 2021 Bis 2025'. Rechtsgutachten im Auftrag der FDP-Bundestagsfraktion. Passau. https://www.fdpbt.de/sites/default/files/2020-06/Rechtsgutachten_BEHG_FDP.pdf.

Wernsmann, Rainer, and Simon Bering. 2020. 'Verfassungsrechtliche Anforderungen an Vorteilsabschöpfungsabgaben'. *Neue Zeitschrift Für Verwaltungsrecht*, no. 8: 497–504.

Cicero. 2021. 'CICERO: Kommuninvest Green Bonds Second Opinion'. https://www.mynewsdesk.com/kommuninvest/documents/cicero-kommuninvest-green-bond-second-opinion-411564.

Klinski, Stefan, Friedhelm Keimeyer, and Sibylle Braungardt. 2021. 'Teilwarmmietemodelle Im Wohnungsmietrecht Als Geeignetes Instrument Zum Klimaschutz? Kurzstudie Zur Rechtlichen Und Praktischen Machbarkeit'. Öko-Institut e.V. https://www.oeko.de/fileadmin/oekodoc/SozUP-Teilwarmmieten.pdf.

Entwurf Eines Gesetzes Zur Änderung Des Gesetzes Über Die Kreditanstalt Für Wiederaufbau Und Weiterer Gesetze, Gesetzesentwurf Der Bundesregierung. 2013. https://dserver.bundestag.de/brd/2013/0215-13.pdf.

Faure, Michael. 2020. 'Die Effektivität Des Umweltrechts Im Zeichen Des Klimawandels'. ZUR, no. 3: 141–50.

Deutscher Bundestag. 2019. Gesetzesentwurf Der Fraktionen CDU/CSU Und SPD, Entwurf Eines Gesetzes Über Einen Nationalen Zertifikatehandel Für Brennstoffemissionen (Brennstoffemissionshandelsgesetz - BEHG).

Bundesregierung. 2019. 'Klimaschutzprogramm 2030 Der Bundesregierung Zur Umsetzung Des Klimaschutzplans 2050'. https://www.bundesregierung.de/resource/blob/974430/1679914/e01d6bd855f09bf05cf7498e06d0a3ff/2019-10-09-klima-massnahmen-data.pdf?download=1

'Wer profitiert vom Berliner Mietmarkt?' 2019. Tagesspiegel. 28 June 2019. https://interaktiv.tagesspiegel.de/lab/mieten-und-renditen/.

Bundesinstitut für Bau-, Staft- und Raumfoschung. 2016. 'Mietrecht Und Energetische Sanierung Im Europäischen Vergleich'. 13. https://www.bbsr.bund.de/BBSR/DE/veroeffentlichungen/bbsr-online/2016/bbsr-online-13-2016-dl.pdf?__blob=publicationFile&v=32016.

Antoni, Johannes, Julia Borger, Michael Kalis, and Simon Schäfer-Stradowski. 2019. 'Verfassungsmäßigkeit Des Entwurfs Zum Brennstoffemissionshandelsgesetz (BEHG-E). Rechtswissenschaftliches Kurzgutachten Und Stellungnahme'. https://www.ikem.de/publikationen/gesamtliste/?tgid=&yr=&type=&usr=&auth=9#tppubs.

Hallof, Irene J. 2013. Das Vermieter-Mieter-Dilemma bei der energetischen Gebäudesanierung: eine rechtliche und ökonomische Analyse. Praxis und Theorie des Bau- und Immobilienrechts 20. Berlin Brüssel: Lexxion Der Juristische Verlag.

Franzius, Claudio. 2018. 'Rechtsprobleme Des Kohleausstiegs'. Neue Zeitschrift Für Verwaltungsrecht, no. 21: 1585–91.

'Just over 5 Million Dwellings in Sweden'. n.d. Statistiska Centralbyrån. Accessed 24 December 2023. https://www.scb.se/en/finding-statistics/statistics-by-subject-area/housing-construction-and-building/housing-construction-and-conversion/dwelling-stock/pong/statistical-news/dwelling-stock-2020-12-31/.

Description of Results

7

In this chapter, the solutions proposed in Chap. 6 will be assessed for their potential transposition into the German legal system. Referring to the Swedish solutions and employing the functional and contextual method of comparative law, the functionality of these solutions within the German legal context will be evaluated. Additionally, the non-legal context of these solutions will also be considered. This approach ensures a comprehensive analysis, taking into account not only the legal frameworks but also the socio-economic and cultural factors that may influence the implementation and effectiveness of the Swedish models in Germany.

7.1 Critical Evaluation of Partial Warm Rent Model

The coalition agreement for 2021–2025 between SPD, Bündnis 90/Die Grünen, and FDP also addresses the concept of partial warm rent, with a commitment to examine it in greater detail.[1] The coalition agreement for 2021–2025 between SPD, Bündnis 90/Die Grünen, and FDP also addresses the concept of partial warm rent, with a commitment to examine it in greater detail.[2] While this chapter will not delve into the legal nature of a

[1] See the whole text in German under: 'Mehr Fortschritt Wagen Bündnis Für Freiheit, Gerechtigkeit Und Nachhaltigkeit Koalitionsvertrag 2021— 2025 Zwischen Der Sozialdemokratischen Partei Deutschlands (Spd), Bündnis 90 / Die Grünen Und Den Freien Demokraten (Fdp)'. n.d. https://www.spd.de/fileadmin/Dokumente/Koalitionsvertrag/Koalitionsvertrag_2021-2025.pdf.

[2] The coalition agreement, 'Mehr Fortschritt Wagen Bündnis Für Freiheit, Gerechtigkeit Und Nachhaltigkeit Koalitionsvertrag 2021— 2025 Zwischen Der Sozialdemokratischen Partei Deutschlands (Spd), Bündnis 90 / Die Grünen Und Den Freien Demokraten (Fdp)'. n.d. https://www.spd.de/fileadmin/Dokumente/Koalitionsvertrag/Koalitionsvertrag_2021-2025.pdf. p. 71.

J. Karras, *Legal Barriers to the Energy Modernisation of Dwellings Occupied by Low-income Tenants and Opportunities to Overcome these Barriers*, Schriftenreihe des Instituts für Klimaschutz, Energie und Mobilität,
https://doi.org/10.1007/978-3-658-44193-7_7

coalition agreement,[3] it is noteworthy that should the government enforce the debt brake, it would pose an additional obstacle to landlords financing energy modernisation.

The concept of partial warm rent itself is not entirely convincing. There is ambiguity surrounding the calculation of heating costs under this system. Landlords face uncertainties, such as the time needed to recoup investments through partial warm rent, and the duration of tenancies. Furthermore, legal concerns about implementing partial warm rent arise, particularly regarding the verification of landlords providing basic heating and tenants bearing costs solely for their heating behaviour beyond the basic provision.

A short study on the legal and practical feasibility of the partial warm rent models in residential tenancy law has calculated an annual net burden for tenants following energy-efficient refurbishment.[4] In all example calculations (considering the status quo, a 50:50 percentage model with a modernisation levy of 8% and 4%) with or without renovation subsidies, tenants incur additional costs. However, burdens are lesser in the 50:50 percentage model with a 4% modernisation levy without renovation subsidies, and in all models with such subsidies.[5] Nevertheless, these models do not cater to low-income tenants lacking financial resources, who will remain unable to afford living in energy-efficient housing due to the additional costs, even if they are relatively lower. For those receiving social assistance under SGB XII, the added costs may not be deemed reasonable if more affordable housing is available in the rental market.

7.2 Reform of Public Law Levies

The reform of public law levies in Germany presents constitutional challenges. Detailed analysis of the legal uncertainties surrounding CO_2 pricing through the national ETS was conducted in Chap. 6. Given the constitutional risks associated with this instrument, it is not advisable to use it for financing the energy modernisation of residential rental properties. Similar legal concerns apply to the building renovation levy proposed by *Hallof*, which was also discussed in the previous chapter. The potential of energy tax reform to address the landlord-tenant dilemma under German law will be explored in the following sections.

There is a strong consensus among scientists on the basic framework of a climate policy-oriented energy taxation with a CO_2 component, directly targeting climate policy goals (the so-called direct tax incentive effect). Taxing energy consumption is effective in promoting reduction in energy use and enhancing energy efficiency. Additionally, the growth of renewable energies could be encouraged through lower tax rates and

[3] See more: Kloepfer 2018, p. 1799f.

[4] See: Klinski/Keimeyer/Braungardt 2021.

[5] Klinski/Keimeyer/Braungardt 2021, p. 16.

exemptions for renewables (the "indirect tax incentive effect").[6] Such measures would positively influence municipal heating planning as a part of the energy transition. Currently, heat planning, and strategic municipal planning are still in their very early stages in Germany.[7] However, an expansion of these practices is anticipated in the near future.

A rational approach to taxation must consider whether different use sectors should be treated uniformly or differently. One argument for differentiation, particularly between the electricity and heating sectors, is the socio-political need to avoid overburdening low-income households.[8] Against this argument speaks that energy poverty can be combated by other instruments that do not set undesirable incentives regarding energy consumption. However, energy poverty can be addressed with other instruments that do not create counterproductive incentives in terms of energy consumption. Yet, uniform treatment of all use sectors under energy tax law is complex due to the varied mix of instruments in different sectors, which collectively impact energy consumption, efficiency, and emissions.[9]

Contrasting with Sweden, where a CO_2 tax serves as an incentive for landlords to undertake energy modernisation of rental properties, Germany could leverage this tax not only as a tool for reducing CO_2 emissions in the residential building sector but also as an additional financial source for energy modernisation of rented dwellings. Taxes aimed at steering behaviour are only problematic under fiscal constitutional law if they create a "strangling effect", which is not generally the case if the tax merely makes a substance more expensive. A strangling effect is presumed only if the tax aims to practically prohibit the taxable event.[10] In the case of the CO_2 component, as long as the energy tax is not set excessively high and does not unduly burden polluters, it should be feasible to use it for taxing energy source consumption in the residential building sector without creating such an effect.

As a purely federal tax, the energy tax generated the highest revenue among all federal taxes in 2021—€37.1 billion.[11] The revenue from the CO_2 component could be utilised as an additional financing source for building renovation by the federal government. This could be achieved by establishing a climate or building modernisation fund, into which these revenues could be channelled. Owners of rental buildings could then access financing through subsidy/loan programmes financed by the energy tax, including

[6] Rodi 2018, p. 743.

[7] Köhler et al., 2021, p. 9.

[8] Gawel/Purkus 2015, p. 77; Rodi 2018, p. 746.

[9] Ibid.

[10] BVerfGE 16, 147 (161); BVerfGE 28, 61 (80); Klinski/Keimeyer 2019, p. 1468.

[11] DeStatis: Steuereinnahmen 2021 summieren sich auf 833 Milliarden Euro, 'Steuereinnahmen 2022 summieren sich auf rund 896 Milliarden Euro'. n.d. Statistisches Bundesamt. Accessed 24 December 2023. https://www.destatis.de/DE/Themen/Staat/Steuern/Steuereinnahmen/steuereinnahmen.html..

its CO_2 component. This approach would mitigate the landlord-tenant dilemma, as tenants would be exempt from the modernisation levy. At the same time, landlords would be relieved to some extent, receiving modernisation subsidies and partial state support. The specific design of such subsidy programmes could further incentivise landlords. This instrument could be even more effective if combined with the partial warm rent model, where landlords already bear part of the heating costs. This arrangement would allow them to directly benefit from energy savings, providing an additional incentive for energy modernisation.

However, the tax burden must correspond to the benefits or costs of public service usage, adhering to the principle of cost-equivalence. According to the equivalence principle of taxation, taxes are viewed as prices for public services. In the case of the CO_2 component of the energy tax, this would relate to climate protection. Yet, if environmental protection is financed by a tax proportional to consumption, this principle is violated, as there is no longer a direct correlation between the distribution of environmental protection benefits and the financing burden.[12]

The ability-to-pay principle in taxation asserts that individuals with equal financial capacity should be taxed equally (horizontal fairness) and those with different capacities should face different taxation levels (vertical fairness). This principle remains valid as long as income is an indicator of the ability to pay. For low-income households, this implies lighter taxation. Introducing a CO_2 component in energy tax, however, would breach the principle of vertical equity, as it would disproportionately burden low-income households relative to their income.[13]

This problem could be partly solved by refunding the CO_2 pricing to the citizens.[14] Nonetheless, this solution raises constitutional concerns. The federal competence for a CO_2 refund could be based on the competencies outlined in Art. 74 (1) No. 11 GG (law of the economy), Art. 74 (1) No. 24 GG (air pollution control) within the concurrent legislation framework, or Art. 105 (2) GG in conjunction with Art. 106 (1) No. 2 GG (consumer taxes).[15] This competence could derive from the so-called competence-establishing genuine connection (*kompetenzbegründender Sachzusammenhang*).[16] In this context, the legislature must clarify that the CO_2 refund's primary purpose is behavioural steering: emitters first pay for the CO_2 intensity of their energy consumption and then receive a refund. For instance, a lump-sum refund would incentivise individuals to reduce their CO_2 emissions.[17] The further constitutional concerns depend on the design

[12] Radulescu 2021, p. 15.

[13] Ibid, p. 16.

[14] Kahl 2021, p. 26.

[15] Kahl 2021, p. 26.

[16] Pieroth, in: Jarass/Pieroth 2020, GG, 15. Auflage 2018, Art. 70, Rn. 7; Kahl 2021, p. 26.

[17] Kahl 2021, p. 26.

of such a refund.[18] They are, however, beside the scope of this thesis and will not be analysed in detail here.

One should note, however, that the tax revenues from the CO_2 component, which are channelled into the aforementioned fund (be it a climate or a building renovation one), and the subsequent subsidising of energy modernisation from this source, cannot be considered a refund. This is because the reduction for socially vulnerable tenants does not seem to be unconditional. It rather still depends on other factors, namely whether the landlord opts to renovate and applies for the subsidy. Those tenants whose landlords decide against such renovation still have to contend with the previous burdens.

In general, tax reform as a mechanism to finance energy modernisation of dwellings occupied by low-income households is viable for both regions—Berlin, as a growing city with a high proportion of rented properties, and Spree-Neiße, a more rural area with lower population density and correspondingly lesser demand for rental housing. The funds are to be allocated to the owners as a non-repayable subsidy, thus absolving them of any risks associated with the modernisation charge. However, tax reform is intrinsically linked to political determination and is time-consuming, as a (federal) law must progress through all parliamentary stages.[19] From a climate protection standpoint, prompt action is essential, yet a multifaceted approach is also required. Hence, additional measures addressing the landlord-tenant dichotomy and focusing on low-income households are necessary.

7.3 Instruments Mobilising Private Investment

To address the energy poverty faced by tenants, it is essential to remedy the market distortions intensified by the legal barriers previously analysed. Sweden's effective practice demonstrates that debt financing, specifically through the issuance of green bonds for energy modernisation, can facilitate the mobilisation of additional investments into residential housing. This thesis will examine whether such a solution is replicable in Germany and, if so, under what conditions.

7.3.1 Municipal Green Bonds

7.3.1.1 Status quo in Germany

The Swedish experience demonstrates that financing complex municipal investment projects through green bonds can act as a significant catalyst for enhanced climate action. In contrast, in Germany, this approach plays a relatively minor role for municipalities.

[18] See more: Kahl 2021, p. 26–29.

[19] Karras/ Novikova 2021, p. 28.

In 2018, approximately 20% of municipal investment was financed through debt capital, with bonds and debt certificates contributing to only about 4%.[20] Until now, German municipalities have primarily financed their investments through more conservative means. Smaller municipalities typically have limited funds, resulting in constrained capabilities for implementing energy transition initiatives at the local level. Conversely, larger municipalities and municipal associations generally have access to more substantial financial resources.[21]

In Sweden, the municipal housing sector accounts for a large share of the rental housing: approximately 20% of the entire housing sector and a half of the rental sector are accounted for by municipal public housing companies.[22] For this reason, the Swedish housing market system is often referred to as a "corporatist housing market system".[23] This distinguishes Sweden from Germany, where the market share of municipal housing in the rental housing market is 10%, although this share varies significantly from region to region.[24]

The role of the municipal housing stock in dealing with different housing policy tasks in Germany is, however, enormous. The Federal Institute for Research on Building, Urban Affairs and Spatial Development (BBSR) conducted research in 2018 to define the role of thereof.[25] A total of 918 local authorities, including 780 cities and municipalities and 138 districts (*Landkreise*), participated in the research.[26] According to the survey, the municipalities see the most significant importance of their housing stocks in providing housing for different population groups: 87% of the respondents view the supply of housing for low-income households as the primary task; 82% of the respondents also see it as necessary to support selected social target groups such as large families or older people with the municipal housing (multiple-choice was allowed). The least importance was assigned to municipal housing in terms of contribution to the municipal budget—only 25% of the respondents considered this task necessary.[27] Thus, municipalities are an essential stakeholder in addressing energy poverty and decarbonising dwellings occupied by low-income tenants.

Moreover, the importance of applying new financial instruments for the energy transition at the municipal level is based primarily on constitutional reasons. Due to the

---

[20] Brand/Steinbrecher 2019, p. 1.

[21] Brand/Steinbrecher 2021, p. 1; Karras, ZKF 3/2022, p. 54.

[22] 'Just over 5 Million Dwellings in Sweden'. n.d. Statistiska Centralbyrån. Accessed 24 December 2023. https://www.scb.se/en/finding-statistics/statistics-by-subject-area/housing-construction-and-building/housing-construction-and-conversion/dwelling-stock/pong/statistical-news/dwelling-stock-2020-12-31/. .

[23] Knutgård/Kristiansen 2013, p. 97; Hananel et al. 2018, p. 462 f.

[24] BBSR 2021, p. 9.

[25] See BBSR 2021.

[26] Ibid, p. 12.

[27] Ibid, p. 78.

separation of budgets, one of the most essential principles of the federal fiscal constitution, which is laid down in Art. 109 (1) GG, German municipalities cannot rely on the federal government to finance the local energy transition. According thereto, the budgets of the Federation and the *Länder* are autonomous and independent of each other.[28] The federal government generally has three options to support the municipalities in financing necessary investments:

- By a one-off contribution to the reduction of old municipal debts;
- By establishing support programmes;
- By establishing a future fund supporting municipal infrastructure and housing companies as a holding company.

The federal government has also made use of these possibilities in recent years. However, they have not been sufficient to compensate for the investment backlog at the municipal level.[29] According to the KfW Municipal Panel 2021, this backlog amounts to €149 billion.[30] German municipalities, however, are relatively conservative regarding the use of innovative financial instruments. Nearly 80% of municipal investments are comprised of their own resources, subsidies from the federal and Länder governments, and purpose-tied investment allocations. Debt financing accounts for only 20% (primarily traditional bank loans).[31]

There are several reasons why German municipalities may not use green bonds as an instrument to finance their energy transition projects, particularly the renovation of the rented dwelling stock. According to KfW economic research on municipal green bonds in Germany,[32] issuing green bonds is viable from a specific volume, which most German municipalities cannot reach. High costs and limited experience in managing capital market financing are the primary obstacles for the municipalities. Numerous documentary obligations are required to issue a green bond. Furthermore, external partners need to be involved in this process. Additional expenditures are required for an external opinion on a green bond and the issuer's framework.[33]

However, as the experience of other countries,[34] especially Sweden, financing conditions can be easily adapted to municipal needs. There is more planning certainty due

[28] Kramer/Hinrichsen/Lauterbach 2012, p. 897; Karras, ZKF 3/2022, p. 54.

[29] Krebs 2020, p. 5f.; Karras, ZKF 3/2022, p. 54 f.

[30] Brand/Steinbrecher, KfW Municipal Panel 2021, p. 1.

[31] Brand/Steinbrecher 2019, p. 2.

[32] Brand/Steinbrecher 2019.

[33] Ibid, p. 3.

[34] Ile-de-France is another municipality that issued green and sustainable bonds for seven categories of projects, i. a. for the construction and renovation of buildings following a sustainable development approach as well as for developing and renovating the social housing stock, see more:

to the interest rate usually being fixed for the entire term. The capital providers are also diversified, making the municipality more independent of individual banks. Additionally, issuing green bonds for financing energy transition projects offers a good marketing opportunity for the municipality to establish its reputation as climate-friendly. Smaller municipalities can also utilise this instrument through coordination and cooperation between several administrative units, which allows them to lower transaction costs, among other benefits.[35]

7.3.1.2 Legal Barriers

However, Germany might face legal barriers that prevent municipalities from issuing green bonds. There are potential barriers at various levels, namely, at the European, national, and Länder levels, which hinder municipalities from utilising this financial instrument to mobilise investment.

European Level

Firstly, the debt brake under European law warrants consideration. At the European level, the relevant debt brake originated in the Treaty on European Union, commonly known as the Maastricht Treaty. This treaty introduced legal norms on the Economic and Monetary Union (Art. 126 (1) TFEU, surveillance procedure according to Art. 121 and 126 TFEU), and was later amended by Protocol No. 12[36] on the excessive deficit procedure. According to Art. 51 TEU, this protocol is an integral part of the Union's primary law. The Stability and Growth Pact, the EU's framework for fiscal and macroeconomic surveillance, supplemented and concretised these treaty norms through the reform of secondary legal acts in 2011.[37]

According to Art. 126 (1) s. 1 TFEU, the Member States are obliged to avoid excessive public deficits. The Commission monitors the development of the budgetary situation and the level of public debt in the Member States. Essential, but not conclusive, criteria for this are laid down in Art. 126 (2) lit. a and b TFEU, in conjunction with Protocol No. 12 on the excessive deficit procedure.[38] "Public" refers to the general government sector, which in Germany's federal structure is subdivided into the federal level, the Länder level, the local (municipal) level, including the associations of municipalities,

'Green and Sustainability Bond Framework of Region Ile-De-France'. n.d. https://www.iledefrance.fr/sites/default/files/medias/2020/06/framework_Region_IdF_en.pdf. .

[35] Brand/Steinbrecher 2019, p. 3.

[36] Protocol (No. 12) on the excessive deficit procedure, Official Journal 115, 09.05.2008 P.0279-0280.

[37] Hermes/Vorwerk/Beckers 2020, p. 4.

[38] Further details are laid down in Regulation (EC) 1479/2009, last amended by Regulation (EU) 220/2014.

and social security funds, excluding commercial transactions.[39] The public deficit is understood as the financial balance of the general government sector, as defined by the European System of National and Regional Accounts (ESA 2010),[40] established by Regulation (EU) No 549/2013 of the European Parliament and of the Council of 21 May 2013. ESA 2010 is "an internationally compatible EU accounting framework for a systematic and detailed description of the economy".[41] This system, primarily developed by Eurostat and the national statistical offices, is designed to meet European needs, defining how economic processes in the EU are to be statistically recorded in the national accounts. Additionally, questions of interpretation regarding the determination of the Maastricht indicators are also addressed by a comprehensive and regularly updated Manual of Government Deficit and Debt (MGDD).[42]

Important indicators for adherence to budgetary discipline and the reference values for fiscal deficit and debt include the ratio between the planned or actual deficit of the public sector and the GDP of the Member State being assessed. The controlled ratio is set at 3% for the deficit to GDP. For the ratio of public debt to GDP, a reference value of 60% applies.[43] Germany's debt-to-GDP ratio was 69.3%,[44] which was 0.5% higher than in 2020. Among other factors, measures taken to address the Covid-19 pandemic have influenced this debt-to-GDP ratio. Considering the statistics from recent years and forecasts for the coming years, a significant and continuous decline appears relatively unlikely. Therefore, compliance with the budgetary discipline pursuant to Art. 126 (2) TFEU is not expected in the immediate future. However, ESA 2010 is pertinent for determining the Maastricht debt level. Not all financial liabilities of the state are included in Maastricht debt. Of the eight instrument categories distinguished by the ESA, only three are relevant for Maastricht debt (liability positions AF.2, AF.3, and AF.4). Accordingly, the Maastricht criteria are not applicable for differentiating between the public and private sectors, as they pertain to those state units that do not primarily produce for the commercial market and are mainly financed by mandatory levies, as well as to those that predominantly redistribute income and assets.[45]

---

[39] Gröpl in: Gröpl 2019, Einleitung, Rn. 74.

[40] Ibid.

[41] https://ec.europa.eu/eurostat/web/esa-2010, accessed on 28.05.2022; Hermes/Vorwerk/Beckers 2020, p. 4.

[42] The last edition from the year 2016 is available under: <https://ec.europa.eu/eurostat/en/web/products-manuals-and-guidelines/-/KS-GQ-16-001>, accessed on 30.05.2022; Deutsche Bundesbank, p. 60.

[43] Häde in: Calliess/Ruffert 2016, Art. 126 AEUV Rn. 23.

[44] https://www.bundesbank.de/de/aufgaben/themen/deutsche-schuldenquote-2021-auf-69-3-prozent-gestiegen-888382, accessed on 28.05.2022.

[45] Deutsche Bundesbank 2018, p. 60; Karras, ZKF 3/2022, p. 56.

All the German administrative units relevant to the Maastricht debt level are divided into core and extra budgets. Extra budgets include all funds, institutions and enterprises that have been separated from the core budgets and are newly established. Most of them perform government tasks, whereas their legal form is irrelevant. As far as the core budgets are concerned, only those units are essential for the ESA, which are subject to direct and indirect government control in their core business and are essentially non-market producers. However, even if the share of market activity predominates, they are still irrelevant for calculating Maastricht debt. Moreover, ESA does not consider all the government financial obligations e.g., shares and units in investment funds are equity instruments and are not subject to Maastricht debt. On the other hand, bonds issued by the federal government are part of debt securities (AF.3) and are thus relevant for Maastricht debt.[46] Considering all those mentioned above, the Maastricht criteria do not forbid the municipalities to issue green bonds, provided they are issued either by the investment funds in which the municipalities hold shares or by the entities acting as market producers.[47]

However, the Member States' obligations arising from the Stability and Growth Pact under EU law have been extended by the Treaty on Stability, Coordination and Governance (TSCG) (the so-called European Fiscal Compact), which entered into force in 2013. According thereto, the Member States commit themselves to realise a balanced budget and to transpose the rules of this treaty into the national law "through provisions of binding force and binding character, preferably constitutional, or otherwise guaranteed to be fully respected and adhered to throughout the national budgetary processes" (Art. 3 (2) s. 1 of the TSCG).[48] Germany has, however transposed these provisions into national law through the ordinary (federal) law—*das Gesetz zur innerstaatlichen Umsetzung des Fiskalvertrags ins nationale Recht vom* 15.07.2013[49]—without any amendments to the Constitutions neither on federal nor on *Länder* levels.

As an outcome of the obligations established in the Fiscal Compact, §51 (2) of the Budget Principles Act (*Haushaltsgrundsätzegesetz,* HGrG) was amended to ensure that the structural overall state financing deficit of the Federation, the Länder, and the municipalities does not exceed an upper limit of 0.5% of nominal GDP. A number of legal acts[50] concretise the monitoring of macroeconomic imbalances.[51] The Fiscal Compact is "only" an international treaty,[52] which established international treaty law standing

[46] Deutsche Bundesbank 2018, p. 60; Karras, ZKF 3/2022, p. 56.

[47] Karras, ZKF 3/2022, p. 56.

[48] Hermes/Vorwerk/Beckers 2020, p. 5.

[49] BGBl I 2013, 2398.

[50] Regulation (EU) No 1176/2011; Regulation (EU) No 1174/2011; Regulation No 1175/2011; Council Directive 2011/85/EU.

[51] Weber 2013, p. 378; Karras, ZKF 3/2022, p. 56 f.

[52] Gröpl in: Gröpl 2019, Einleitung, Rn. 100; Karras, ZKF 3/2022, p. 57.

alongside EU primary law. According to the BVerfG, the debt criteria of the European Fiscal Compact do not create any new obligations for Germany, as its actual contents are in line with constitutional and EU law requirements.[53]

The question of whether the ESA applies to the debt brake of the Federation or the *Länder* can be found in the wording of Regulation No. 549/2013 itself. According to Art. 1 (4), this Regulation does not oblige any Member State to use ESA 2010 in compiling accounts for its own purposes. The historical interpretation of Art. 109 (3) GG also does not provide any indications that the national debt brake should follow the EU rules.[54] However, according to Art. 109a (2) s. 2 GG, which is relatively new as it was introduced in the context of amendments to the fiscal constitution in 2017, the Stability Council must consider the requirements and procedures from legal acts based on the TFEU for compliance with budgetary discipline when monitoring compliance with the debt brake by the Federation and the *Länder*. Nevertheless, this provision should not be understood as suggesting that Regulation (EU) No. 549/2013 is decisive for the question of the allocation of legally independent units within the framework of the debt brake, according to Art. 109 (3) GG by virtue of national constitutional law. Art. 109a (2) GG regulates only the monitoring process and has no influence on the material rules of Art. 109 (3) GG. The monitoring thereof is the point.[55] The Fiscal Compact is also not covered by the reference of Art. 109 (2) GG, as it was not adopted based on the TFEU but on international law, and not all Member States of the Union are party to it.[56]

Overall, EU law does not generally prohibit municipalities from issuing green bonds. However, there are specific considerations to be taken into account, such as the issuance by investment funds in which the municipalities hold shares, or by entities acting as market producers. Nevertheless, these considerations are not so significant as to render the issuance of green bonds by German municipalities uneconomical.[57]

Federal Level

With the so-called *Föderalismusreform II* (Federal Reform II), the GG introduced the debt brake. The core of these constitutional amendments was the introduction of a general ban on balancing budgets with revenues from loans. According to Art. 109 (3) s. 1 GG, this is stipulated for both the Federation and the *Länder*. Art. 115 (2) s. 1 GG almost identically articulates this for the federal budget—balancing the budget through

[53] BVerfGE 145, 317 (432 f.); Gröpl in: Gröpl 2019, Europarechtlichen Vorgaben und Bezüge, Rn. 99; Ebenda: Einleitung, Rn. 100;

[54] Hermes/Vorwerk/Beckers 2020, p. 24; more to the origin and amendments of Art. 109 GG see: Heun in: Dreier 2015, Art. 109 Rn. 2 et seq.

[55] Hermes/Vorwerk/Beckers 2020, p. 24.

[56] Heun in: Dreier 2015, Art. 109a, Rn. 37; Kube in: Dürig/Herzog/Scholz 2021, Art. 109a GG; Rn. 26f.

[57] Karras, ZKF 3/2022, p. 57.

new debts is generally forbidden. Previously, according to the old version of Art. 115 (1) s. 2 GG, the revenue from loans was allowed to cover fully the sum of the budgeted expenditure for investment, even in a standard case. The investment-related debt limit of the GG did not apply to the *Länder*.[58] The debt brake of the GG was an absolute novelty in German constitutional history. While the limitation of borrowing by the federal government was not new to the GG, there had never been such a comprehensive prohibition before. Additionally, there is an abundance of other principles concretising the federal structure, which together define the legal framework for the budgetary system of the Federation and the *Länder*.[59]

This reform supplemented the Art. 109 GG with a net borrowing restriction binding for the Federation and the *Länder*, which was independent of the European Stability and Growth Pact. At the same time, the amended provision should be understood in the light of the obligations arising from the Pact mentioned above. Thus, whether the constitutional debt brake introduced by the *Föderalismusreform II* prohibits issuing green bonds by German municipalities remains questionable.[60] Moreover, the newly introduced general prohibition of structural new indebtedness shall be understood as a mandatory requirement, with only a few exceptions.[61] Accordingly, a deficit of up to 0.35% of GDP is acceptable for the Federation. At the same time, the *Länder* are bound to zero limits without exception under Art. 109 (3) GG. In this context, credits include financial flows from loans,[62] other similar contracts, and government bonds raised on the capital market.[63] New financial instruments such as leasing or forfeiting may also be considered credit instruments within the meaning of Art. 109 (3) GG.[64]

The GG provides for some exemptions from the general prohibition on debt financing.[65] According to Art. 109 (3) s. 2 Hs. 2, the Federation and the *Länder* may provide regulations for symmetrical consideration of the effects of an economic development deviating from the normal situation in the upswing and downswing, as well as an exemption regulation for natural disasters or extraordinary emergencies that are beyond the control of the state and significantly affect the state's financial situation. Art. 115 (2) s. 3 contains similar wording and stipulates that if the economic development deviates from the normal situation, the effects on the budget in the upswing and downswing must be considered symmetrically. The procedural issues are laid down in Art. 115 (3) s. 4

[58] Lenz/Burgbacher 2009, p. 2562.

[59] Kramer/Hinrichsen/Lauterbach 2012, p. 897.

[60] Kube in: Dürig/Herzog/Scholz 2021, Art. 109 GG, Rn. 25 f.

[61] Heun in: Dreier 2015, Art. 115 GG Rn. 22; Karas, ZKF 3/2022, p. 57.

[62] Heun in: Dreier 2015, Art. 115 GG Rn. 22.

[63] Kube in: Maunz/Dürig 2021, Art. 109 GG Rn. 134; Kramer/Hinrichsen/Lauterbach 2012, p. 898.

[64] Kube in: Maunz/Dürig 2021, Art. 109 GG Rn. 137.

[65] More information to this matter: Siekmann in Sachs et al. 2021, Art, 109 GG Rn. 70ff: Bundesrat,- Drucksache 262/09, p. 22f.

and s. 5. S. 6 of this legal norm says that in the event of natural disasters or exceptional emergencies beyond the control of the state and significantly affecting the state's financial position, these borrowing limits may be exceeded based on a resolution passed by a majority of the members of the Bundestag. No other exceptions to the prohibition of borrowing are possible, not even under the aspect of the unity of the constitution.[66]

The debt brake of the GG is significantly stricter than the deficit rules of Art. 126 TFEU and the European Stability Pact. The EU provisions implement the fiscal concept of fiscal sustainability; according thereto, the growth rate of the public debt shall correspond to the growth of production potential. At the same time, zero debt is not required thereby. Nevertheless, the debt ratio is stable in a growing economy.[67]

Although the GG does foresee some exceptions from the ban on new debt, these do not extend to balancing revenues and expenditures of energy transition projects with revenues from loans. Consequently, the constitutional debt brake presents an obstacle to the issuance of green bonds by the *Länder* and possibly by the federal government. However, whether these provisions also apply to the budgets of municipalities and municipal associations remains a matter for debate.[68]

The wording of Art. 109 (3) GG is formulated open-ended,[69] and a grammatical interpretation does not clearly reveal whether the legislator intended to include municipalities and their associations within its scope. A systematic interpretation of the finance constitution demonstrates that the legislature explicitly mentions municipalities (and municipal associations) in several instances (Art. 104c GG, 105 (3) GG, 108 (7) GG), even when discussing the financing of the *Länder* (e.g., Art. 104b (1) GG, 104d GG). Therefore, it can be inferred that the absence of municipalities in the wording of Art. 109 (3) is intentional, not an oversight, indicating a deliberate decision to exclude them.[70]

The historical interpretation of Art. 109 (3) GG also supports the argument against including municipalities within the scope of this provision. The original version of Art. 109 GG focused solely on the separation and autonomy of the budgetary management of the Federation and the *Länder*, identical in wording to the current paragraph 1. The concept of separate financial management for the Federation and the *Länder* was initially presented in Art. 37 of the Report on the Constitutional Convention on Herrenchiemsee, albeit in a general competence context. The autonomous budget management of municipalities and their associations was also perceived as being within the scope of this regulation.[71] When adopting the provision as Art. 121, the Parliamentary Council decided against a separate inclusion of municipalities. This provision was later repositioned to

[66] Lenz/ Burgbacher 2009, p. 2562.

[67] Wieland in: Herdegen et al. 2021, §24 Rn. 39.

[68] Karas, ZKF 3/2022, p. 57.

[69] Reimer in: BeckOK, Art. 109 Rn. 52a; Siekmann in: Sachs et al. 2021, Art. 109 Rn. 85.

[70] Karas, ZKF 3/2022, p. 57.

[71] Rodi in: Dolzer/Vogel/Graßhof, 2015, Art. 109, Rn. 5f.; Heun in: Dreier 2015, Art. 109 Rn. 2.

the beginning of the section on finance as Art. 123a, before the budget law. The regulation of municipal finance was thus left to the legislation of the *Länder*.[72]

Moreover, the draft of the act amending the Basic Law (Art. 91c, 91d, 104b, 109, 109a, 115, 143d) suggests that including deficits of social security funds and municipalities in the budget drafting process was not originally intended.[73] Indeed, this would impose unrealistic information requirements on the budget drafting of the Federation and the *Länder*, both in terms of content and timing.[74] However, this argument could be contested. Currently, all *Länder* have comprehensive prohibitions on indebtedness under municipal law. The municipal supervisory authorities could set deadlines for municipal budget drafting to reduce information requirements.[75]

According to a minority opinion in the literature, the extension of the scope of Art. 109 (3) GG to municipalities is justified by the provision's background under European law. Initially, the European debt limit of 0.5% of GDP was to be incorporated into the GG. The intention was to set a limit of 0.15% of nominal GDP for the *Länder*'s debt in addition to the 0.35% limit for the federal government's cyclical debt. However, this idea was abandoned during the discussions of the Federalism Commission II due to the failure to agree on the horizontal distribution of the 0.15% limit. These two amounts were meant to correspond to the European debt limit. According to this minor opinion, the limitation of 0.5% of GDP laid down in Art. 3 (1) of the Fiscal Compact, which was transposed into national law, applies to all public and subsidiary budgets. As a result, the separation of the municipal budgets was not intended from the outset.[76] However, this view overlooks the Fiscal Compact's nature as an international treaty standing alongside EU primary law. The norms of the Fiscal Compact have been transposed into national law by ordinary federal act. A broader interpretation of the constitution here would contradict the legislator's intention, who had several opportunities to include municipalities in the wording of the norm, as the fiscal constitution has been amended several times in recent years.

An additional argument of this minority opinion regarding the scope of Art. 109 (3) GG in relation to municipal budgets is based on teleological interpretation. This interpretation views the municipalities as part of the Länder, arguing that without such consideration, there is a risk of burdens being shifted from *Länder* budgets to the municipal level. Consequently, seeing municipalities as addressees of Art. 109 (3) GG follows the legal principle of protecting against circumvention and establishes practical concordance with Art. 28 (2) s. 3 GG.[77]

---

[72] Heun in: Dreier 2015, Art. 109, Rn. 2.

[73] BBundesrat, Drucksache 262/09.

[74] Bundesrat, Drucksache 262/09, p. 22; Reimer in Beck, Art. 109 Rn. 51.

[75] Reimer in Beck, Art. 109 Rn. 52b.

[76] Ibid, Rn. 52 c.

[77] Reimer in BeckOK, Art. 109, Rn. 52 d.

The majority view in legal literature rejects the inclusion of municipalities in the scope of Art. 109 (3) GG, using the same arguments previously mentioned.[78] The BVerfG has already left the direct binding of municipalities and associations of municipalities under the old version of Art. 109 (2) GG open. It stated: "It can be left open whether and, if so, to what extent the municipalities can be bound by Art. 109 II GG in their budgetary measures and the statute-maker can be bound by Art. 109 (2) GG in fiscal regulations [...]."[79] A concrete ruling of the BVerfG on this matter is still missing.[80]

The arguments for including municipalities in the provision above are not persuasive. Affirming such a binding effect would contradict the legislator's intent, who purposefully omitted mention of municipalities in Art. 109 (3) GG yet referred to them in other sections of the fiscal constitution. Considering that the fiscal constitution has been amended multiple times since the *Föderalismusreform II*,[81] the legislator could have revised Art. 109 (3) GG to include municipalities in its scope if that had been the intention. Furthermore, another binding interpretation of this provision by the BVerfG is also absent.[82]

However, a clear federal constitutional debt brake prohibiting municipalities from issuing green bonds is not present in the German legal system, leading to legal uncertainties for municipalities. From a legal policy perspective, it would be beneficial to introduce a precise regulation that unambiguously allows for credit financing of the energy transition at all levels, including municipalities.

Länder Level

Constitutional Law of Länder

Considering that the federal constitution does not explicitly prohibit municipalities from issuing green bonds, such a prohibition could still be derived from the constitutions of the *Länder*. However, for the purposes of this thesis, only Berlin and the district of Spree-Neiße in Brandenburg have been chosen as examples for detailed consideration, with a brief analysis of the legislation of other *Länder* also included.

Under Art. 109 (3) s. 5 GG, the *Länder* have the authority to regulate the detailed arrangements for their budgets within their constitutional competencies. These regulations must comply with the general debt brake, which prohibits loan revenue,

[78] Siekmann in: Sachs et al. 2021, Art. 109 GG Rn. 85; Kube in Maunz/ Dürig 2021, Art. 109 GG Rn. 119; Karas, ZKF 3/2022, p. 57.

[79] BVerfG NVwZ 1990, 356 (357).

[80] Karas, ZKF 3/2022, p. 57.

[81] See more: Dreier in: Dreier 2015, GG, Änderungen des Grundgesetzes.

[82] Karas, ZKF 3/2022, p. 58.

as outlined in Art. 109 (3) s. 1 GG. While the *Länder* are bound by this fundamental debt brake, they also retain considerable leeway in structuring their budgets.[83] A critical aspect of this provision is its potential restriction on the budgetary autonomy of the *Länder*, as expressed in Art. 109 (1) GG, potentially impacting their statehood (*Eigenstaatlichkeit*).[84] The unique feature of the German Federation is that not only the state as a whole but also its member states—the *Länder*—possess state quality. They have their constitutions and the right to articulate their fundamental state norms independently.[85]

The constitutional autonomy of the *Länder* is, however, closely linked with the framework of the federal structure. Despite their statehood, the sovereignty of the *Länder* is limited as they are not entitled to all the features of a sovereign state, e.g. to form their armed forces, introduce their currency, have a monopoly on taxation, use force as well as to generally decide about their subject status under international law.[86] Yet, financial sovereignty remains a key competency for the *Länder*, ensuring their ability to organise and fulfil their responsibilities autonomously.[87] Sufficient financial resources are considered essential to the core of their statehood.[88] Fiscal sovereignty, which encompasses both expenditure and revenue, is crucial for the *Länder*'s autonomy. The GG in Art. 109 (1) and Art. 104a (1) reinforces this by stipulating that the Federation and the *Länder* are autonomous and independent in their budgetary management and bear separate expenses arising from their duties.[89]

However, Art. 109 (4) GG presents an exception to this principle of separation. It authorises the Federation to establish budgetary law and administration principles applicable to both the Federation and the *Länder*, albeit limited to principles and comprehensive regulations that do not restrict the involved addressees' leeway. These principles must be uniformly applied to both levels. The Federation has exercised this competence through the enactment of the Budget Principles Act and the Act on Promotion of Stability and Growth of the Economy (StabG).[90] Overall, the general intent of the GG is to maintain separate budgetary systems for both levels.

However, it remains unclear whether the credit autonomy of the *Länder* constitutes an essential part of their budget autonomy. As previously mentioned, the original 1949 version of the GG did not include a debt brake. The Parliamentary Council reached a

[83] Lenz/Burgbacher 2009, p. 2564.

[84] Dreier in: Dreier 2015, Art. 28 GG, Rn. 42.

[85] BVerfGE 36, 342 (360 f.); Dreier in: Dreier 2015, Art. 28 GG, Rn. 42.

[86] Dreier in: Dreier 2015, Art. 28 GG, Rn. 43 f.; Sommermann in: Mangoldt/ Klein/Starck 2018, Art. 20 GG, Rn. 26 f.

[87] Fassbender 2009, p. 739.

[88] Kramer/Hinrichsen/Lauterbach 2012, p. 900, more to this matter: Fassbender 2009, p. 739.

[89] Fassbender 2009, p. 739.

[90] Kramer/Hinrichsen/Lauterbach 2012, p. 897.

consensus that budgetary separation is a fundamental aspect of federalism and that fiscal sovereignty should permit the Länder to incur state debts. This right was further reinforced by the constitutional amendment of 1967. According to the previous version of Art. 109 (4) s. 1 GG, the *Länder*'s right to incur state debts could only be restricted under federal law to prevent a disturbance of the overall economic equilibrium. Therefore, the legislator aimed to formally and substantively separate the budgetary systems of the *Länder* and the Federation.[91]

The concept of the tax state, as established in the fiscal constitution, posits that the state should primarily finance itself through taxes. In this context, a pertinent issue arises: if tax resources are insufficient to cover all financial needs of the *Länder* and loan financing is prohibited, such a general debt brake might infringe upon Art. 79 (3) GG, particularly concerning the guarantee of the basic substance of the Länder's sovereignty.[92] However, Art. 105-107 GG comprehensively and conclusively regulate the distribution of tax revenue between the Federation and the *Länder,* ensuring an adequate allocation of financial resources. If the *Länder* still face financial shortfalls, solutions other than debt financing, such as tax reforms, can be sought. Therefore, credit autonomy is not deemed essential for the sovereignty of the *Länder*. Furthermore, they are bound by a duty of loyalty to the Federation, an unwritten constitutional principle, which is aligned with the federal state principle of Art. 20 (1) GG and is also part of the core area protected by Art. 79 (3) GG. Given the Federation's obligations under the Stability Compact, it is in line with loyalty to the Federation to restrict the budget autonomy of the *Länder*.[93] Moreover, the legislator has allowed the *Länder* some degree of discretion in Art. 109 (3) s. 5 GG, ensuring a minimum level of autonomy. According to this provision, the *Länder* are entitled to regulate exceptions provided for in the GG independently.[94]

Summarising the aforementioned points, it is clear that the *Länder* are bound by the federal constitutional debt brake. Nonetheless, they retain some discretion in regulating exceptions to this rule and its application, which will be further analysed. A key question arises concerning the effect of regulations transposed into the *Länder* constitutions on the ability of municipalities to issue green bonds.

At this point, it should first be distinguished between the structure of the Federal Republic of Germany under state organisation law and the structure under administrative law. The former structure is based on the concept of government at two levels—the federal and the *Länder*.[95] Municipalities (and their associations) are granted legal status

[91] Ibid, p. 900.

[92] See: Dreier in: Dreier 2015, Art. 79 Rn. 48; Fassbender 2009, p. 40 f.

[93] Kramer/Hinrichsen/Lauterbach 2012, p. 900 f.

[94] Ibid, p. 901.

[95] Groh 2010, p. 2; Karras, ZKF 4/2022, p. 80.

under Art. 28 (2) of the GG, which they can enforce in court against other levels of government. The GG considers them as part of a multi-level state structure. The municipalities also exercise state power within the meaning of Art. 20 (2) s. 1 GG when performing self-government tasks. Thus, they must be legitimised by the people, and the chain of legitimation must be traceable. Municipal self-government is, therefore, considered an element in the democratic structure of the state. However, in terms of state organisational law, considering a two-tier federal-state concept, the municipalities are incorporated into the functional context of the *Länder* government.[96] The competencies under the federal constitution are also allocated to the Federation and the *Länder*. However, the administrative structure of Germany is three-tiered, and a municipal level is thereby a level of a so-called indirect state administration (*mittelbare Staatsverwaltung*) besides the federal and the *Länder* level.[97] The municipalities are integrated into the state as administrative authorities.[98] The fiscal constitution is based on the two-tier state structure and considers municipalities as part of the *Länder*,[99] which is particularly evident in Art. 106 (9), 107 (2) GG. Moreover, the *Länder* have final decision-making competence regarding the actual financial resources of their municipalities.[100] At the same time, the GG endows the municipalities with special rights in the form of the guarantee of self-government, distinguishing them from the general *Länd*er organisation. For example, according to Art. 28 (2) s. 3 GG is the basis of their financial self-responsibility.[101]

As analysed above, the *Länder* have some discretion concerning the regulations of the constitutional debt brake due to their constitutional autonomy.[102] Following *Bravidor's*[103] categorisation of the *Länder*'s constitutional provisions concerning the debt brake and considering the amendments thereto as of 1st June 2022, all 16 constitutions of the *Länder* can be divided into three groups:

- Constitutions with a far-reaching debt brake, which not only prohibit debt financing but also specify circumstances where revenue from credits is accounted to the Länder, even if the subject to debt financing is a legal entity (Bremen and Rhineland-Palatinate),

[96] Mehde in: Dürig/Herzog/Scholz 2021, Art. 28 GG Rn. 11.

[97] Groh 2010, p. 2; Karras, ZKF 4/2022, p. 80.

[98] Groh 2010, p. 2; Gawert, p. 27 f.; Karras, ZKF 4/2022, p. 80.

[99] Seiler in: Dürig/Herzog/Scholz 2021, Art. 106 GG; Rn. 163; Huber in: Mangoldt/Klein/Starck 2018, Art. 107 GG Rn. 41.

[100] Groh 2010, p. 2; Bravidor 2016, p. 74; Karras, ZKF 4/2022, p. 80;

[101] Seiler in: Dürig/Herzog/Scholz 2021, Art. 106 GG; Rn. 163. In the city-states Berlin and Hamburg, which, unlike Bremen, have no municipalities, the municipal revenue responsibilities are established in Art. 106 (5), Art. 106 (7) GG is due to the Land, partly directly, partly by analogy with Art. 106 (6) s. 3 GG, see: Seiler in: Dürig/Herzog/Scholz 2021, Art. 106 GG; Rn. 166.

[102] Wieland in Herdegen et al. 2021, §24 Rn. 49; Fassbender 2009, p. 741.

[103] Bravidor 2016, p. 185f.

- Constitutions with a simple debt brake, where exceptions are only permitted in cases of natural disasters or extraordinary emergencies (Bavaria, Brandenburg, Hamburg, Hesse, Mecklenburg-Western Pomerania, Lower Saxony, Saxony, Saxony-Anhalt and Schleswig-Holstein),
- Constitutions that do not regulate the debt brake (Baden-Württemberg, Berlin, North Rhine-Westphalia, Saarland and Thuringia).

Only two constitutions explicitly refer to municipal self-government (Art. 141 (2) in conjunction with Art. 137 (5) of Constitution of Hesse and Art. 95 (3) in conjunction with Art. 85, 87 of Constitution of Saxony.[104] According to these provisions, the debt brake does not affect those clauses that provide for burden and financial equalisation in the implementation of own and transferred tasks.[105]

Brandenburg, encompassing the district of Spree-Neiße, has implemented a simple debt brake regulation in its constitution, as outlined in Art. 103 of the Constitution of Brandenburg. In contrast, Berlin's constitution does not address issues of debt financing. However, it remains unclear whether the constitutional debt brakes in those constitutions that include debt financing provisions actually prohibit municipalities from issuing green bonds.

The wording of the constitutional debt brake provisions across the *Länder* is generally similar. Typically, the *Länder* budgets are to be balanced without resorting to debt financing. Exceptions to this rule are permitted in cases of economic developments deviating from the norm or in events of natural disasters or extraordinary emergencies beyond the state's control that significantly impact the state's financial situation.[106]

In the context of the extension of the debt brake under *Länder* constitutional law, municipalities present the same arguments as those mentioned in relation to the federal constitutional debt brake. The notable exception here is the Land of Bremen. According to Art. 143 (1) and Art. 146 (1) of the BremVerf, in conjunction with Art. 131a (1) of the BremvVerf, the cities of Bremen and Bremerhaven each constitute a municipality within the state of Bremen, and thus, the debt brake provision is applicable to the municipalities' financial systems. The legislators in other *Länder* had the opportunity to extend the debt brake to municipalities through the enactment or amendment of their constitutions (given that the constitutions of 14 Länder were enacted post the GG) but, to date, only Bremen has availed this option.[107]

---

[104] Those Länder, whose constitutions have no regulation to the debt brake, have regulated this matter in their ordinary Land law (mostly in the budget regulations), see more: Karras, ZKF 4/2022, p. 80 f.

[105] Bravidor 2016, p. 323.; Karras, ZKF 4/2022, p. 81.

[106] E.g., Art. 82 of Constitution of Bavaria, Art. 117 of Constitution of Rheinland-Pfalz, Art. 65 of Constitution of Mecklenburg-Western Pomerania.

[107] The same extension of the debt brake to the municipalities also took place in the Constitution of Bremen of 1947. Nevertheless, Bremen is a special case concerning its structure.

No other case law from the constitutional courts of the *Länder* has deemed municipalities subject to the debt brake under Land constitutional law. Nevertheless, neither the federal government nor the *Länder* are fundamentally barred from introducing municipal debt brakes in the Federal constitution or in the constitutions of the *Länder*, respectively.[108]

Ordinary Legislation of Länder

Although the debt brake provisions incorporated into the constitutions of certain *Länder* do not explicitly prohibit municipalities from issuing green bonds, some *Länder* have regulated the debt brake in their ordinary legislation.[109] Generally, these financial regulations do not include municipalities within their regulatory scope. However, exceptions exist for the city-states (*Stadtländer*) of Berlin and Hamburg, which, instead of municipalities, have districts. This is referenced in Art. 4 (1) Verf Bln and Art. 4 (1), (2) HmbVerf. Notably, in Hamburg, the constitution explicitly states that state and municipal activities are not separate (Art. 4 I HmbVerf), meaning the constitutional debt brake under Art. 72 (1) HmbVerf applies to the districts of Hamburg. In Berlin, where the debt brake is not regulated in the constitution but in the Financial Regulation of Berlin (LHO Bln), the debt brake from ordinary Berlin legislation applies to its districts, as per §18 LHO Bln.[110] For Bremen, the constitutional debt brake is applicable only to the city of Bremen and not to Bremerhaven, reflecting the unique structure of this two-city state.

Overall, the constitutional debt brake is explicitly applicable in only two of the sixteen Länder: Bremen and Hamburg. Berlin, while not regulating the debt brake in its constitution, does so in §18 LHO Bln of its Financial Regulation. Given Berlin's structure, which consists of districts rather than municipalities, the issue of issuing municipal green bonds is not relevant here. According to §18 (1) s. 1 LHO Bln, the budget should generally be balanced without revenues from loans, although sentence 2 allows for exceptions. The remaining *Länder* have addressed the issue of municipal debt brakes within their municipal regulations,[111] which form part of the *Länders'* ordinary legislation.[112]

The analysis of the ordinary legislation across various *Länder* has demonstrated that they permit their municipalities to take out loans for investments and investment promotion measures. Additionally, these municipalities are allowed to reschedule loans for similar purposes if alternative financing is either unavailable or economically

---

[108] Karras, ZKF 4/2022, p. 81. See more: Bravidor 2016, p. 324f.

[109] Bravidor 2016, p. 182f.; Karras, ZKF 4/2022, p. 81.

[110] Karras, ZKF 4/2022, p. 81.

[111] In some Länder they are called *Gemeindeordnung,* in some Länder—*Kommunalverfassung*, in Saarland—*Kommunalselbstverwaltungsgesetz*, in Thüringen—*Kommunalordnung.*

[112] Karras, ZKF 4/2022, p. 81 f.

unreasonable.[113] Notably, Hesse stands out as the only *Land* that authorises its municipalities to obtain investment loans even when alternative financing options are available, as specified in §103 HGO.[114]

It follows from the wording of the corresponding Länder provisions that the legal possibility for municipalities to take out credit for investment financing is subsidiary.[115] Municipalities must first ensure that no alternative financing options are available. If alternatives exist, the municipality must then conduct a scrutiny test to determine whether such alternatives are economically reasonable. This evaluation considers both the specific financial situation of the municipality and its overall economic circumstances, potentially determined by profitability or comparative calculations.[116]

Municipalities have a degree of discretion in deciding whether other financing alternatives would be economically reasonable. This discretion arises from the need for evaluative considerations and forecasts in assessing economic feasibility.[117] Issuing green bonds to finance energy modernisation of the residential building sector and other energy transition projects offers clear benefits for municipalities, including an enhanced image as a reliable debtor, potentially leading to lower interest rates compared to private investors. Financing with external capital is economically reasonable as long as the costs of this capital remain below the profits that can be generated with it.[118]

However, in legal practice, credit financing may be deemed economically unreasonable if low-interest loans from subsidy programmes,[119] such as those offered by the *Kreditanstalt für Wiederaufbau* (German Reconstruction Loan Corporation, KfW).[120] The Bavarian State Chancellery, for instance, in its Amendments to the Notice on Municipal Lending,[121] stipulates that when deciding on credit borrowing, municipalities must ensure that the terms are acceptable and market-aligned. This involves obtaining

---

[113] Cf. Art. 71 (1) in conjunction with Art. 62 (3) BayGO; §87 (1) in conjunction with § 78 (3) BWGemO; §74 (1) BbgKVerf in conjunction with §64 (3) BbgKVerf; §52 (1) in conjunction with § 44 (3) KV M-V; §120 (1) NKomVG in conjunction with §111 (6) NKomVG; §86 (1) GO NRW in conjunction with §77 (4) GO NRW; §103 (1) RhPfGemO in conjunction with §94 (4) RhPfGemO; §92 (1) KSVG in conjunction with §83 (2) KSVG; §82 (1) SächsGemO in conjunction with §73 (4) SächsGemO; §108 (1) KVG LSA in conjunction with §99 (5) KVG LSA; §85 (1) GO SH in conjunction with §76 (3) GO SH; §63 (1) ThürKO in conjunction with §54 (3) ThürKO; §103 (7) HGO.

[114] Karras, ZKF 4/2022, p. 81.

[115] Sedlmaier in: BeckOK Kommunalrecht BayGO, §62 Rn. 14.

[116] Brüggen 2004, p. 7; Karras, ZKF 4/2022, p. 81.

[117] VGH München BeckRS 2012, 50788, Rn. 31; Sedlmaier in: BeckOK Kommunalrecht BayGO, §62 Rn. 15.

[118] Wieland in: Herdegen et al. 2021, §24 Rn. 44.

[119] Cf. Sedlmaier in: BeckOK Kommunalrecht BayGO, §62 Rn. 16.

[120] Glinder/ Friedl 2011, p. 50; Karras, ZKF 4/2022, p. 82.

[121] Cf. Bayerisches Ministerblatt, 2019 Nr. 346.

and comparing several loan offers, also in relation to the municipality's budgetary situation, unless credits from development banks (e.g., KfW) are used on behalf of the state.[122] Therefore, credits from development banks are privileged as they do not require comparison with other instruments.

For municipalities facing informational barriers regarding innovative financial instruments, opting for simpler procedures with lower legal risks, such as those offered by development banks, may be more favourable than issuing green bonds. Furthermore, KfW offers various subsidy programmes for municipalities, including those for energy modernisation of residential buildings, renewable energy support, waste disposal, environmentally friendly technologies, and climate-friendly mobility.[123] Consequently, municipalities may be restricted in using debt financing instruments like green bonds in these areas, as such financing might be considered economically unreasonable.[124]

Some *Länder* also distinguishes between the credits and the so-called credit-like instruments, the latter usually being permissible for municipalities, albeit requiring approval from the supervisory authority.[125] Laws in those *Länder* that make this distinction do not define what credit-like instruments are. However, in practice, this term is understood to include investment financing instruments common in the private sector,[126] such as green bonds. Credit-like instruments are also applicable when other financial alternatives are either unavailable or not economically viable.[127]

Consequently, the current legal situation in most Länder inhibits the issuance of green bonds for financing energy modernisation of the building stock without considering the profitability of such projects.[128] Municipalities face a significant risk when issuing green bonds for the energy modernisation of residential buildings, as they are effectively bound by the "economic reasonability" of KfW programmes, especially considering their low interest rates. Legislators recognise the need to mobilise private investments, but the availability of (state) funds from KfW, an institution under public law as per § 1 (1) KredAnstWiAG, obstructs this path. Moreover, the Federal Government's liability

[122] Bayerisches Ministerblatt 2019 Nr. 346, p. 3.

[123] Cf. 'Förderprodukte für energieeffiziente Sanierung – Übersicht | KfW'. n.d. Accessed 24 December 2023. https://www.kfw.de/inlandsfoerderung/Privatpersonen/Bestandsimmobilie/Energieeffizient-Sanieren/Förderprodukte/.

[124] Glinder/Friedl 2011, p. 50; Karras, ZKF 4/2022, p. 82.

[125] Cf. Art. 72 (1) BayGO; §87 (5) BWGemO; §74 (5) BbgKVerf; §120 (6) NKomVG; §86 (4) NKomVG; §82 (5) SächsGemO; §108 (6) KVG LSA; §85 (5) GO SH; §64 ThürKO; cf. Henkes in: BeckOK Kommunalrecht B-W GemO, §87 Rn. 44.

[126] Cf. 'Kommunales Wirtschaftsrecht - Rechtsgeschäfte - Kreditähnliche Rechtsgeschäfte | Kreditähnliche Rechtsgeschäfte'. n.d. Accessed 24 December 2023. https://www.lds.sachsen.de/kommunal21/?ID=144&art_param=26.

[127] Cf. .

[128] Wieland 2016, p. 479; Karras, ZKF 4/2022, p. 81.

for loans provided by KfW, as stated in §1a KredWiAG, and KfW's role in financing numerous state tasks on behalf of the state, according to §2 (1) KredWiAG, further complicates matters. The current investment backlog in the municipal sector clearly indicates that the state alone cannot shoulder the financial requirements for the energy transition and essential climate protection measures. Without alternative financing solutions, there is a risk that Germany will fail to meet its climate targets.[129]

Additionally, a strained market situation in the construction sector is leading to significant price increases for investments. In 2019, the average price increase for public buildings was 6%. In some regions, these increases are so substantial that they negate the benefits of low-interest rates.[130] This trend particularly impacts municipalities, which are responsible for a large portion of construction investment and often have comparatively low debt levels, thereby not benefiting as much from low-interest rates.[131] Therefore, promoting additional financing instruments among municipalities is crucial.

7.3.1.3 Possible Solutions under the Current Legal Framework

There are numerous legal uncertainties currently surrounding the issuance of municipal green bonds. In some Länder, the issuance is already rendered impossible due to their unique structural frameworks. In others, the legislation neither explicitly prohibits nor permits such issuance, creating a realm of uncertainty, particularly for smaller municipalities. This ambiguity often leads them to opt for more conventional financial instruments. An adjustment of the legal framework and the removal of existing barriers are highly desirable from a legal policy perspective. Such changes are crucial to meet set climate targets and successfully decarbonise the residential building sector. However, adapting the legal situation to facilitate this change is a time-consuming process, especially considering the intricacies of the legislative procedure.[132] Despite these challenges, some viable options within the current legal framework are conceivable and are discussed in the following sections.

Issuance of Green Bonds by Legally Autonomous Legal Entities

According to the prevailing opinion in the literature, the constitutional debt brake does not encompass independent legal persons, irrespective of whether they are established under public or private law. This interpretation holds true even if the Federal

[129] Karras, ZKF 4/2022, p. 82.

[130] Statistisches Bundesamt 2021:; Brand/Salzgeber 2019, p. 4.

[131] Brand/Salzgeber 2019, p. 4.

[132] For example, according to Art. 125 (3) BremVerf, a resolution to amend the BremVerf (Constitution of Bremen), other than by referendum, can only be passed if the Bürgerschaft votes by a majority of two-thirds of its members. According to Art. 125 (2) s. 1 BremVerf, three readings of a constitutional amendment must take place on different days.

Government bears liability for their debts, regardless of the legal basis. For municipalities looking to issue green bonds legally, one conforming option could be to delegate this task to an entity not subject to the debt brake.[133] Such independent legal entities, possessing their own budgets, are not considered part of the Federation under budgetary law, even if they fall under the indirect administration of the federal government. The same principle applies if a *Land* finances such a legal entity or is liable for its obligations, whether through explicit guarantees or institutional liability.[134]

Case law also supports this approach. As per the Berlin Constitutional Court, a Land has the discretion under budgetary constitutional law to divest portions of its public tasks and financial management, transferring them to independent legal persons under public or private law that it controls. While there is a risk of circumventing constitutional regulation on the debt brake, the current wording of the constitutional provisions regulating the debt brake does not allow for extending its scope to borrowing by public legal persons that undertake their material tasks and finance loans in their own names. There are no general criteria to provide legal certainty in distinguishing between permissible borrowing and circumvention of the budgetary constitutional provisions. Therefore, the motive behind each borrowing instance must be scrutinised by constitutional courts on a case-by-case basis.[135]

As previously examined, some Länder have implemented more stringent debt brake regulations in their Land Constitutions, explicitly prohibiting debt financing by legal entities. For instance, Bremen forbids such debt financing by legal entities when the Land can directly or indirectly exercise a controlling influence over them, owing to ownership, financial participation, articles of association, or other provisions (cf. Art. 131a (5) BremVerf). Most Länder have not taken the step to regulate the debt brake more strictly than stipulated in the GG.[136]

Therefore, municipalities could establish independent legal entities capable of issuing green bonds on their behalf. The nature of these entities, whether under private or public law, is not a critical factor. The crucial aspect is that the municipality assigns them material tasks and the borrowing transactions are conducted in their names. Essentially, these legal entities would be responsible for fulfilling financial tasks such as issuing green bonds, alongside other duties such as executing infrastructure or climate mitigation projects within the municipality.[137] This approach could be particularly beneficial for small

[133] Siekmann in: Sachs et al. 2021, Art. 115 GG, Rn. 7; Hermes/ Vorwerk/ Beckers 2020; Heun in: Dreier 2015, Grundgesetz-Kommentar. Art. 115 GG Rn. 24; Kube in: Maunz/ Dürig 2021, Grundgesetz-Kommentar, Art. 115 GG, Rn. 60; Heintzen in: v. Münch et al. 2021, Art. 115 GG, Rn. 12;

[134] Siekmann in: Sachs et al. 2021, Art. 115 GG, Rn. 71.

[135] BerlVerfGH NVwZ-RR 2003, 537 (540); Siekmann in: Sachs et al. 2021, Art. 115 GG, Rn. 75.

[136] Karras, ZKF 4/2022, p. 83.

[137] Heun in: Dreier 2015, Grundgesetz-Kommentar, Art. 115 GG Rn. 24; Hermes/ Vorwerk/ Beckers 2020, S. iv; Karras, ZKF 4/2022, p. 83.

municipalities with limited financing options, enabling them to fund decarbonisation projects through green bonds. They could do this by setting up such a legal entity in collaboration with several municipalities, thereby pooling resources and expertise.

Issuance of Green Bonds by Private Credit Institutions

This instrument, though not new in Germany, has been used irregularly by some banks. An additional viable option for financing the decarbonisation of the residential building stock could be to encourage private credit institutions to issue green bonds for municipal projects. This approach could significantly mobilise private investment. Further details on how this can be executed in a legally compliant manner are provided below.[138]

For private credit institutes, this represents an excellent opportunity to enhance their reputation and attract both existing and new customers. Furthermore, participants in the green bond market could expand their value chain and product portfolios by introducing a green segment on trading venues.[139] This also has the potential to inform and influence those not directly involved in the green bond market.

For instance, Berlin Hyp AG, an independent subsidiary of Landesbank Berlin Holding AG, has issued green and sustainability-linked bonds to refinance commercial real estate in Berlin. The bank developed its own Sustainable Agenda and Sustainability-Linked Bond Framework, focusing on the building sector in alignment with Germany's climate neutrality goals.[140] They have also published their methodology for calculating the carbon footprint, enabling the transparent presentation of CO_2 emissions for their entire loan portfolio. This methodology, developed using a comprehensive bank database, has been verified for its robustness and the relevance of its key performance indicators for the banking and real estate sector by ISS ESG[141]—institutional shareholder service group of companies enabling investors to develop and integrate responsible investing policies and practices.[142]

However, to engage more private law credit institutions in issuing green bonds for municipal building projects, additional financial incentives are needed. Municipalities often face limitations in terms of finances and legal barriers. For instance, the municipal laws of most *Länder* either prohibit or restrict municipalities from providing securities on behalf of third parties.[143] Such security could minimise risks for participants

---

[138] Karras, ZKF 4/2022, p. 83.

[139] Eisinger et al. 2017, p. 14.

[140] ISS ESG 2022.

[141] ISS ESG 2022.

[142] 'About ISS I ISS'. n.d. Accessed 24 December 2023. https://www.issgovernance.com/about/about-iss/#1574276741161-7ca718d3-32ae.

[143] Cf. §88 (1) BWGemO; §75 (1) BgbKVerf; §104 (1) HGO; §121 (1) NKomVG; §87 (1) GO NRW; §57 (1) KV M-V; §104 (1)) RhPfGemO; §93 (1) KSVG; §83 (1) SächsGemO; §108 (7) KVG LSA; §86 (1) GO S-H; §63 (6) ThürKO;

in energy modernisation projects. Although there are exceptions in individual cases to this general prohibition in municipal law, the process is typically time-consuming and fraught with risks, potentially deterring credit institutions from involvement.

At this point, municipalities could assign the task to the aforementioned independent legal entities, established by them, to assume securities for municipal energy transition projects in relation to both the credit institute and investors. The debt brake typically does not apply to the assumption of guarantees by such entities unless the *Länder* has specifically legislated otherwise. This approach aligns with the concept of segregating the budgets of these entities from the municipalities' budgets under budgetary law. In the event of insolvency of a municipal investment project for which a private law credit institute has issued a green bond, the independent legal entity would compensate the credit institute for any asset losses incurred. Furthermore, it would assume security responsibilities vis-á-vis investors, guaranteeing to fulfil the financial obligations of the credit institute in case it fails to do so independently. This arrangement would considerably diminish the risks associated with municipal projects financed via green bonds, thereby encouraging more private entities to participate in funding energy modernisation projects, and consequently, mobilising more private investment.[144]

Through this model, municipalities effectively delegate their energy transition and climate mitigation projects to legally independent entities and private law credit institutions. The advantage here is that even if the delegated project becomes insolvent, and the municipality is indirectly liable for related financial burdens, it could still continue the project at its own expense, thus fulfilling its climate mitigation obligations.[145]

7.3.2 Summarising

German municipalities in almost all *Länder* encounter legal barriers to issuing green bonds on their own. Hessen is a notable exception, as it does not impose as stringent a debt brake on its municipalities compared to other *Länder*. Consequently, Hessen's municipalities could serve as a laboratory for issuing municipal green bonds under existing legislation. However, a broader legal policy amendment is required to explicitly empower municipalities to utilise credit-like financing mechanisms for energy transition projects, given their pivotal role in implementing these transitions on the ground. To circumvent time loss, the legally conforming methods for issuing municipal green bonds (as well as corporate green bonds on behalf of municipalities) have been explored.

[144] Karras/Novikova 2021, p. 29; ESRa P4.3., p. 29; Karras, ZKF 4/2022, p. 83.

[145] Karras/Novikova 2021, p. 29; Karras, ZKF 4/2022, p. 83.

The predominant legislative opinion, supported by some case law, indicates that independent legal entities are not barred by the debt brake from using debt financing, except where the constitutions of certain Länder (e.g., Bremen) explicitly dictate otherwise. In such scenarios, the debt is attributed to the entity that established it, thus invoking the debt brake. Establishing these entities demands careful consideration to ensure they are not created merely to evade legal constraints but are genuinely engaged in tasks that justify debt financing. In practice, these entities issue not 'municipal' green bonds, as their activities are not accounted for by the municipalities, but rather 'corporate' bonds for municipal financing projects. This approach represents a modification of the instrument within the German legal order.

The alternative option, involving existing private law credit institutes to issue green bonds, expands the range of actors working on behalf of municipalities to mobilise private investment. Germany's tradition of state-financed projects through KfW demonstrates the efficiency of such an approach. Nonetheless, this model cannot be the sole means for financing the energy transition, as the state cannot bear the entire financial burden.

The selected regions for comparison, Berlin and the Spree-Neiße district, have distinct legal frameworks. Berlin, being both a city-*Land* and the capital of the Federal Republic of Germany, faces additional legal hurdles in issuing green bonds compared to the municipalities of Spree-Neiße. However, the proposed solutions to these obstacles are applicable to both regions. Notably, case law from the Constitutional Court of Berlin affirms that the debt brake does not apply to independent legal entities, adding legal certainty for this *Land*.

The both modalities of issuing green bonds on behalf of municipalities are suitable for both Berlin and Spree-Neiße. This approach is especially advantageous for smaller municipalities like those in Spree-Neiße, which often have limited access to financial resources. These municipalities can collaborate to establish a joint independent legal entity responsible for carrying out energy transition projects, including issuing green bonds for their financing.

References

'About ISS | ISS'. n.d. Accessed 24 December 2023. https://www.issgovernance.com/about/about-iss/#1574276741161-7ca718d3-32ae.

'Änderung Der Bekanntmachung Über Das Kreditwesen Der Kommunen. Bekanntmachung Des Bayerischen Staatsministeriums Des Innern, Für Sport Und Integration'. 2019. Bayerisches Ministerblatt 2019 346. https://www.verkuendung-bayern.de/files/baymbl/2019/346/baymbl-2019-346.pdf.

Bäcker, Matthias, Pascale Cancik, Wolfgang Durner, Martin Eifert, Peter M. Huber, Matthias Jestaedt, Jens Kersten, et al. 2021. *Handbuch des Verfassungsrechts: Darstellung in transnationaler Perspektive*. Edited by Matthias Herdegen, Johannes Masing, Ralf Poscher, and Klaus Ferdinand Gärditz. München: C.H. Beck.

'BeckOK KommR Bayern | Gemeindeordnung Für Den Freistaat Bayern – Beck-Online'. n. d. Accessed 24 December 2023. https://beck-online.beck.de/?vpath=bibdata%2Fkomm%2FBeck-OKKommRBay_1%2FBAYGO%2Fcont%2FBECKOKKOMMRBAY%2EBAYGO%2Ehtm.

Brand, Stephan, and Johannes Salzgeber. 2019. 'Green Bonds – a Sustainable Alternative for Municipal Infrastructure Finance?' https://doi.org/10.13140/RG.2.2.14972.80005.

Brand, Stephan, and Johannes Salzgeber. 2021. 'Finanzierung Öffentlicher Investitionen: Kredite Allein Helfen Den Kommunen Nicht'. 360. KfW. https://www.researchgate.net/publication/357271249_Finanzierung_offentlicher_Investitionen_Kredite_allein_helfen_den_Kommunen_nicht?_tp=eyJjb250ZXh0Ijp7ImZpcnN0UGFnZSI6InByb2ZpbGUiLCJwYWdlIjoicHJvZmlsZSJ9fQ.

Brand, Stephan, and Johannes Steinbrecher. 2021. 'Too Soon to Sound the All-Clear-Coronavirus May yet Hit Municipal Finances Where It Hurts. KfW Municipal Pannel 2021'. KfW Research. KfW. https://www.kfw.de/PDF/Download-Center/Konzernthemen/Research/PDF-Dokumente-KfW-Kommunalpanel/KfW-Kommunalpanel-2021_KF_EN.pdf.

Bravidor, Christoph. 2016. *Die Vereinbarkeit Der Schuldenbegrenzungsregelungen Mit Der Garantie Der Kommunalen Selbstverwaltung*. Vol. 1315. Schriften Zum Öffentlichen Recht. Duncker & Humblot. https://www.duncker-humblot.de/buch/die-vereinbarkeit-der-schuldenbegrenzungsregelungen-mit-der-garantie-der-kommunalen-selbstverwaltung-9783428146956/?page_id=1.

Brüggen, Georg. 2004. 'Kommentar Zur Sächsischen Gemeindeordnung, §73 Grundsätze Der Einnahmebeschaffung. Stand 01.01.2004'. http://brueggen-ra.de/pdf/KommWirtschaftsrecht/SaechsGemO_p73.pdf.

Bundesinstitut für Bau-, Staft- und Raumfoschung. 2021. 'Kommunale Wohnungsbestände: Mietengestaltung – Ausweitung – Investitionen. Ergebnisse Der BBSR-Kommunalbefragung 2018'. Bonn. https://www.bbsr.bund.de/BBSR/DE/veroeffentlichungen/sonderveroeffentlichungen/2021/kommunale-wohnungsbestaende-dl.pdf?__blob=publicationFile&v=2.

BVerfG, 15.12.1989 – 2 BvR 436/88 – Regelung des Steuermaßstabs für Zweitwohnungssteuer. 1989, 1990 NVwZ 356–57. BVerfG.

BVerfG, 18.02.1970 – 2 BvR 746/68 – Leserbrief. 1970. Beschl. BVerfG.

BVerfG, 22.05.1963 – 1 BvR 78/56 – Werkfernverkehr. 1963. Urt. BVerfG.

BVerfG, 23.05.2017 – 2 BvR 883/14, 2 BvR 905/14 – Ostbesoldung Besoldungsangleichung. 2017. Beschl. BVerfG.

BVerfG, 29.01.1974 – 2 BvN 1/69 – Landesgrundrechte, Niedersächsisches Landesbesoldungsgesetz. 1974. Beschl. BVerfG.

Calliess, Christian, and Matthias Ruffert. 2016. *EUV/AEUV. Das Verfassungsrecht Der Europäischen Union Mit Europäischer Grundrechtecharta*. 5th ed. C. H. Beck. https://beck-online.beck.de/?vpath=bibdata/komm/CalliesRuffertEUVEGV_4/cont/CalliesRuffertEUVEGV.htm.

Council Directive 2011/85/Eu of 8 November 2011 on Requirements for Budgetary Frameworks of the Member States. n.d. Accessed 24 December 2023. https://eur-lex.europa.eu/legal-content/GA/TXT/?uri=CELEX%3A32011L0085.

Council Regulation (EC) No 479/2009 of 25 May 2009 on the Application of the Protocol on the Excessive Deficit Procedure Annexed to the Treaty Establishing the European Community (Codified Version). 2009. *OJ L*. Vol. 145. http://data.europa.eu/eli/reg/2009/479/oj/eng.

Deutsche Bundesbank. 2018. 'Die Maastricht-Schulden: Methodische Grundlagen Sowie Die Ermittlung Und Entwicklung in Deutschland'. Monatsbericht. https://www.bundesbank.de/resource/blob/724596/3f48b7be00f9574cc930717eb5dda090/mL/2018-04-maastricht-data.pdf.

'Deutsche Schuldenquote 2021 auf 69,3 Prozent gestiegen'. n. d. Accessed 24 December 2023. https://www.bundesbank.de/de/aufgaben/themen/deutsche-schuldenquote-2021-auf-69-3-prozent-gestiegen-888382.

Dolzer, Rudolf, Karin Graßhof, and Klaus Vogel. 2004. *Bonner Kommentar Zum Grundgesetz*.

Dürig, Günter, Roman Herzog, and Theodor Maunz, eds. 2021. *Grundgesetz: Kommentar*. München: Beck.

Eisinger, Frederik, Annica Cochu, and Rainer Agster. 2017. 'Der Deutsche Green Bond Markt – Für Ein Langfristiges Wachstum, Studie in Kooperation Mit Dem Bundesverband Öffentlicher Banken Deutschlands'. https://www.adelphi.de/en/system/files/mediathek/bilder/Der%20deutsche%20Green%20Bond%20Markt%20–%20für%20ein%20langfristiges%20Wachstum%20-%20adelphi.pdf.

Epping, Volker, and Christian Hillgruber. 2022. 'Beck'scher Online-Kommentar Zum Grundgesetz – Beck-Online'. 15 February 2022. https://beck-online.beck.de/Dokument?vpath=bibdata%2Fkomm%2Fbeckok_30_bandverfr%2Fcont%2Fbeckok.htm.

'EUR-Lex – 12008M/PRO/12 – EN'. n.d. Text/html; charset=UTF-8. Official Journal 115 , 09/05/2008 P. 0279 – 0280; OPOCE. Accessed 24 December 2023. https://eur-lex.europa.eu/LexUriServ/LexUriServ.do?uri=CELEX%3A12008M%2FPRO%2F12%3AEN%3AHTML.

Fassbender, Bardo. 2009. 'Eigenstaatlichkeit Und Verschuldungsfähigkeit Der LänderVerfassungsrechtliche Grenzen Der Einführung Einer „Schuldenbremse" Für Die Länder'. *Neue Zeitschrift Für Verwaltungsrecht*, no. 12: 737–41.

'Förderprodukte für Energie und Umwelt | KfW'. n.d. Accessed 24 December 2023. https://www.kfw.de/inlandsfoerderung/Öffentliche-Einrichtungen/Kommunen/Energie-und-Umwelt/.

Gawel, Erik, and Alexandra Purkus. 2015. 'Die Rolle von Energie- und Strombesteuerung im Kontext der Energiewende'. *Zeitschrift für Energiewirtschaft : ZfE*, Zeitschrift für Energiewirtschaft : ZfE. – Wiesbaden : Springer Vieweg, Springer Fachmedien Wiesbaden, ISSN 0343-5377, ZDB-ID 132891-8. – Vol. 39.2015, 2, p. 77–103, 39 (2).

'Gesetz Zur Innerstaatlichen Umsetzung Des Fiskalvertrags'. 2013. *Bundesgesetzblatt Teil I*, no. 38 (July): 2398.

Gesetzesantrag Der Länder Baden-Württemberg, Bremen: Entwurf Eines Gesetzes Zur Änderung Des Grundgesetzes (Artikel 91c, 91d, 104b, 109, 109a, 115, 143d). 2009. https://www.bundesrat.de/SharedDocs/drucksachen/2009/0201-0300/262-09.pdf;jsessionid=240F809F37B0451C3BC3AEB715FF5FF8.1_cid374?__blob=publicationFile&v=3.

Glinder, Peter, and Eric Friedl. 2011. *Gemeindehaushaltsrecht Baden-Württemberg: Neues Kommunales Haushalts- und Rechnungswesen mit Kassen- und Prüfungsrecht*. Stuttgart: Kohlhammer.

'Green and Sustainability Bond Framework of Region Ile-De-France'. n.d. https://www.iledefrance.fr/sites/default/files/medias/2020/06/framework_Region_IdF_en.pdf.

Groh, Katrin. 2010. 'Schuldenbremse Und Kommunale Selbstverwaltungsgarantie'. *Landes- Und Kommunalverwaltung*, 1–8.

Gröpl, Christoph, ed. 2019. *Bundeshaushaltsordnung, Landeshaushaltsordnungen: Staatliches Haushaltsrecht: Kommentar*. 2. Auflage. München: C.H.Beck.

Dreier, Horst. 2015. Grundgesetz-Kommentar. https://buch-sauermann.de/shop/item/9783161504945/grundgesetz-kommentar-2-von-horst-dreier-leinen.

Hermes, Georg, Lukas Vorwerk, and Thorsten Beckers. 2020. 'Die Schuldenbremse Des Bundes Und Die Möglichkeit Der Kreditfinanzierung von Investitionen – Rechtslage, Ökonomische Beurteilung Und Handlungsempfehlungen'. Hans Böckler Stiftung. https://www.boeckler.de/de/faust-detail.htm?sync_id=9095.

ISS ESG. 2022. 'Re-Verification of the Sustainability Quality of the Issuer and Green Bond Asset Pool Berlin Hyp AG',. Second Party Opinion. https://www.berlinhyp.de/en/investors/

green-bonds?file=files/media/corporate/investoren/green-bonds/green-bonds/externe-beurteilung/28-03-22-spo-update-berlin-hyp.pdf.

Jarass, Hans-D., Martin Kment, and Bodo Pieroth. 2020. *Grundgesetz für die Bundesrepublik Deutschland: Kommentar*. 16. Auflage 2020. München: C.H. Beck.

'Just over 5 Million Dwellings in Sweden'. n.d. Statistiska Centralbyrån. Accessed 24 December 2023. https://www.scb.se/en/finding-statistics/statistics-by-subject-area/housing-construction-and-building/housing-construction-and-conversion/dwelling-stock/pong/statistical-news/dwelling-stock-2020-12-31/.

Kahl, Hartmut. 2021. 'CO_2-Preis Mit Rückerstattung: (Wie) Geht Das?' *Ifo Schnelldienst*, no. 06: 26–29.

Karras, Jana. 2022a. 'Kommunale Green Bonds Im Lichte Der Schuldenbremse – Teil 1'. *ZKF*, no. 3: 54–58.

Karras, Jana. 2022b. 'Kommunale Green Bonds Im Lichte Der Schuldenbremse – Teil 2'. *ZKF*, no. 4: 80–84.

Klinski, Stefan, and Friedhelm Keimeyer. 2019. 'Zur Finanzverfassungsrechtlichen Zulässigkeit Eines Nationalen Zertifikathandels Für CO_2-Emissionen Aus Kraft- Und Heizstoffen, Rechtswissenschaftliches Kurzgutachten'. https://www.oeko.de/fileadmin/oekodoc/Verfassungsrecht_Emissionshandel_Gebaeude-Verkehr.pdf.

Klinski, Stefan, Friedhelm Keimeyer, and Sibylle Braungardt. 2021. 'Teilwarmmietemodelle Im Wohnungsmietrecht Als Geeignetes Instrument Zum Klimaschutz? Kurzstudie Zur Rechtlichen Und Praktischen Machbarkeit'. Öko-Institut e.V. https://www.oeko.de/fileadmin/oekodoc/SozUP-Teilwarmmieten.pdf.

Kloepfer, Michael. 2018. 'Koalitionsvereinbarungen – Unverbindlich, Aber Rechtlich Relevant'. *Neue Juristische Wochenschrift*, no. 25: 1799–1803.

Knutagård, Marcus, and Arne Kristiansen. 2013. 'Not by the Book: The Emergence and Translation of Housing First in Sweden Introduction'. *European Journal of Homelessness* 7 (January): 93–115.

Köhler, Benjamin, Veit Bürger, Roman Weidinger, Hannes Doreder, Simon Schäfer-Stradowsky, and Dennis Tänzler. 2021. 'Strategische Kommunale Wärmeplanung.' Ariadne-Analysw. Potsdam. https://ariadneprojekt.de/publikation/analyse-strategische-kommunale-waermeplanung/.

'Kommunales Wirtschaftsrecht – Rechtsgeschäfte – Kreditähnliche Rechtsgeschäfte | Kreditähnliche Rechtsgeschäfte'. n.d. Accessed 24 December 2023. https://www.lds.sachsen.de/kommunal21/?ID=144&art_param=26.

Kramer, Urs, Tim Hinrichsen, and Theresa Lauterbach. 2012. 'Die Schuldenbremse Des Grundgesetzes'. *Juristische Schulung*, no. 10: 896–902.

Krebs, Tom. 2020. 'Öffentliche Investitionen: Bedarfe Und Finanzierung. Schriftliche Stellungnahme, Für Die Anhörung Im Haushaltsausschuss Des Deutschen Bundestages Zu Den Anträgen Der Fraktionen DIE LINKE, FDP Und BÜNDNIS90/DIE GRÜNEN (Drucksachen 19/14375, 19/14424, 19/15919, 19/16831 Und 19/16841) Zum Thema Schuldenbremse Und Investitionen'. https://www.bundestag.de/resource/blob/684410/143685b2edde41b8bb29106f3a4b6dbf/Prof-Tom-Krebs-data.pdf.

Lenz, Christofer, and Ernst Burgbacher. 2009. 'Die Neue Schuldenbremse Im Grundgesetz'. *Neue Juristische Wochenschrift*, no. 35: 2561–67.

Mangoldt, Hermann, Friedrich Klein, and Christian Starck. 2018. *Grundgesetz. Kommentar*. 7. München: C. H. Beck. https://beck-online.beck.de/Dokument?vpath=bibdata%2Fkomm%2FMangoldtKlStKoGG%2Fcont%2FMangoldtKlStKoGG.htm&anchor=Y-400-W-MANGOLDTKLSTKOGG&jumpType=Jump&jumpWords=Mangoldt%252fKlein%252f-Stark&readable=Suche%2Bnach%2BKommentar%253a%2BMangoldt.

'Mehr Fortschritt Wagen Bündnis Für Freiheit, Gerechtigkeit Und Nachhaltigkeit Koalitionsvertrag 2021— 2025 Zwischen Der Sozialdemokratischen Partei Deutschlands (Spd), Bündnis 90 / Die Grünen Und Den Freien Demokraten (Fdp)'. n.d. https://www.spd.de/fileadmin/Dokumente/Koalitionsvertrag/Koalitionsvertrag_2021-2025.pdf.

Münch, Ingo von, Philip Kunig, Jörn Axel Kämmerer, and Markus Kotzur, eds. 2021. *Grundgesetz Kommentar*. 7., Neubearbeitete Auflage. München: C.H. Beck.

Novikova, Aleksandra, Tim Langenhorst, and Jana Karras. n. d. 'Systematisierter Und Kategorisierter Juristischer Instrumentenkanon, Energiewende Im Sozialen Raum'. Produkt No. 4.3. https://esra-projekt.de/sites/default/files/2022-02/2021.09.14_esra_0.pdf.

'Overview – ESA 2010 – Eurostat'. n.d. Accessed 24 December 2023. https://ec.europa.eu/eurostat/web/esa-2010.

Radulescu, Doina. 2021. 'Verteilungswirkungen von CO_2-Preisen Und Des Ausbaus Erneuerbarer Energien'. *Ifo Schnelldienst* 6 (2021): 14–18.

Rodi, Michael. 2018. 'Die Zukunft Der Energiesteuern Im Rahmen Der Energiewende'. IKEM. https://www.ikem.de/wp-content/uploads/2018/08/GS_Joecks_Rodi.pdf.

Sachs, Michael, Ulrich Battis, Herbert Bethge, Ralf Brinktrine, Christian von Coelln, and Peter M. Huber. 2021. *Grundgesetz: Kommentar*. 9. Auflage. München: C.H. Beck.

'Steuereinnahmen 2022 summieren sich auf rund 896 Milliarden Euro'. n.d. Statistisches Bundesamt. Accessed 24 December 2023. https://www.destatis.de/DE/Themen/Staat/Steuern/Steuereinnahmen/steuereinnahmen.html.

VGH München, 21.03.2012 – 4 B 11.221. 2012. Urt. VGH München.

Weber, Albrecht. 2013. 'Europa- Und Völkerrechtliche Elemente Der Gewährleistung von Haushaltsdisziplin in Der Währungsunion'. *Europarecht*, no. 4: 375–89.

Wieland, Joachim. 2016. 'Verfassungsrang Für Nachhaltigkeit'. *ZUR*, no. 9: 473–83.

Karras, Jana, and Aleksandra Novikova. 2021. 'Germany. Legal Barriers to the Decarbonisation of Dwellings Occupied by Low-Income Tenants and Financial Incentives to Overcome These Barriers'. IKEM Climate Finance Week. https://www.ikem.de/wp-content/uploads/2021/05/Brief_5_Legal-barriers-to-the-decarbonisation-of-dwellings.pdf.

Hananel, Ravit, Sharon Krefetz, and Ami Vatury. 2018. 'Public Housing Matters: Public Housing Policy in Sweden, the United States, and Israel'. Journal of Planning Education and Research 41 (August): 0739456X1879370. https://doi.org/10.1177/0739456X18793702.

Commission Regulation (EU) No 220/2014 of 7 March 2014 Amending Council Regulation (EC) No 479/2009 as Regards References to the European System of National and Regional Accounts in the European Union. 2014. OJ L. Vol. 069. http://data.europa.eu/eli/reg/2014/220/oj/eng.

Statistisches Bundesamt. 2021. 'Preisindizes Für Die Bauwirtschaft, 2. Vierteljahresausgabe'. Fschserie 17. Reihe 4.

Epping, Volker, and Christian Hillgruber. 2022. 'Beck'scher Online-Kommentar Zum Grundgesetz - Beck-Online'. 15 February 2022. https://beck-online.beck.de/Dokument?vpath=bibdata%2Fkomm%2Fbeckok_30_bandverfr%2Fcont%2Fbeckok.htm.

Dietlein, Johannes, and Arne Pautsch. n.d. 'BeckOK KommunalR BW - Beck-Online'. Accessed 1 December 2023. https://beck-online.beck.de/?vpath=bibdata%2fkomm%2fBeckOKKommRB-W_17%2fcont%2fBECKOKKOMMRBW.htm.

Maunz, Theodor, and Günter Dürig. 2021. Grundgesetz Kommentar. 94. Lieferung. München

8 Conclusion

Energy transition affects almost every sector of the economy, with the residential building sector being a significant contributor to overall CO_2 emissions. Therefore, decarbonising this sector is imperative. Concurrently, it is vital to ensure that the energy transition does not disproportionately burden low-income households, thereby advocating for a just transition. Low-income tenants, often a more vulnerable social group, typically reside in energy-inefficient housing due to affordability constraints, which precludes them from accessing better living standards.

Adopting a multi-methodological approach, this research addresses the legal barriers confronting the residential building sector inhabited by low-income tenants. Initially, it delves into the background of this demographic, frequently associated with poverty and energy poverty. Consequently, these two phenomena are scrutinised within the German and Swedish contexts, offering insights into the challenges and potential solutions for ensuring an equitable energy transition.

Germany currently does not address energy poverty within its legislative framework. Furthermore, the German government asserts that energy poverty is non-existent in the country, attributing this to Germany's welfare status. Despite this official stance, the phenomenon of energy poverty does exist, albeit in a latent form that often escapes the attention of policymakers. In response to this oversight, a specific definition of energy poverty suitable for the German context has been developed. According to this definition, *energy poverty is characterised by the inability to either keep the house adequately warm or to pay utility bills in the last 12 months, or both; or reliance on social (publicly funded) assistance that covers energy bills to a reasonable extent, coupled with the necessity to conserve energy to avoid exceeding the amount covered by public funds; or challenges in maintaining adequate warmth in the household without sacrificing other*

J. Karras, *Legal Barriers to the Energy Modernisation of Dwellings Occupied by Low-income Tenants and Opportunities to Overcome these Barriers*, Schriftenreihe des Instituts für Klimaschutz, Energie und Mobilität,
https://doi.org/10.1007/978-3-658-44193-7_8

needs, alongside difficulties in making ends meet, even when the individual does not receive (publicly funded) social aid.[1] This comprehensive definition facilitates the examination of barriers faced by a broader social group, including those who do not benefit from social aid but still struggle to pay their energy bills by sacrificing other essential needs.

The status of tenure and the type of dwelling play a pivotal role in addressing the challenges of energy poverty. In the current geopolitical climate, with rising energy prices, the economic importance of energy savings has dramatically increased. Decarbonising residential stock is crucial for energy savings, making it more critical than ever. In Germany, and particularly in Brandenburg and Berlin, rented dwellings constitute a significant portion of the residential stock, accounting for over half and over 80% respectively.[2] The process of decarbonising rented properties is often hindered by the landlord-tenant dilemma, where neither party has sufficient incentives to invest in thermal efficiency improvements. Tenants, not being property owners, typically lack the motivation to finance energy modernisation as they cannot fully reap its benefits. Conversely, landlords, despite having the right to increase rent following energy renovations, must shoulder the high upfront costs and capital tie-up.

For low-income tenants, this dilemma is exacerbated, as they generally lack access to financial resources, even if they are willing to invest in energy modernisation. A further complication is the 'reverse split incentive', which arises when occupants are not responsible for their utility bills, such as when these are covered by the state, leading to little or no interest in energy conservation.

The objectives of this research included identifying the legal barriers to energy modernisation in the context of low-income occupants of rented dwellings and exploring potential solutions informed by literature analysis and the Swedish experience. Initially, the research examined relevant EU policies. The EU's regulatory framework in the realm of dwelling energy efficiency is intricate and has undergone multiple revisions over the past decade to align with the Union's ambitious climate protection and energy goals. The foundational regulatory policies are typically enacted as directives, granting member states discretion in their transposition. Implementing acts are often issued as regulations, requiring no national legislation transposition. While the legislation of member states in this field is somewhat harmonised, they retain considerable leeway in transposition measures and are encouraged to pursue more stringent CO_2 reduction targets and measures, provided they do not contravene EU law. Therefore, analysing the legal and non-legal contexts of Sweden offers valuable insights into potential solutions that could be applicable to Germany.

---

[1] The definition has been developed by the author, drawing upon the discussion and findings mentioned in Sect. 3.1.1.

[2] 'Wer profitiert vom Berliner Mietmarkt?' 2019. Tagesspiegel. 28 June 2019.https://interaktiv.tagesspiegel.de/lab/mieten-und-renditen/.

Employing the context method of comparative law as a key instrument in this thesis necessitated an in-depth examination not only of legal systems and their approaches to the topic at hand but also of the broader context surrounding these laws. For the German analysis, two contrasting regions were selected: Berlin, a thriving metropolitan city benefiting from the energy transition, and the Spree-Neiße district, a rural area largely dependent on conventional energy sources like coal. At first glance, these regions appear markedly different, particularly in their challenges related to decarbonising dwellings occupied by low-income households. In Berlin, approximately 83% of dwellings are rented,[3] compared to around 50% in Brandenburg.[4] Despite these differences, both regions grapple with a common issue: financing the decarbonisation of dwellings, particularly for socially vulnerable groups who lack financial assets and cannot bear the costs associated with energy modernisation.

The legal analysis of German legislation revealed that the German legislator has only guaranteed a subsistence minimum upholding human dignity under Art. 1 (1) GG, in conjunction with the welfare state principle under Art. 20 (1) GG. This minimum is primarily defined in social legislation and does not extend to ensuring a specific level of thermal efficiency in housing. Low-income households are only afforded limited security in terms of adequate housing: state-subsidised rent should be reasonable, with the definition of 'reasonable' left to the Länder. Social courts scrutinise whether the state is not burdened with unreasonable costs. Subsidised dwellings are compared within the lower rent price segment of a spatial district, with energy efficiency often not factored into the determination of reasonable costs. In Germany, low-income tenants also lack entitlement to demand energetic modernisation of their dwellings.

German tenancy law incorporates a significant social component, protecting tenants from rapid rent increases or rental agreement terminations due to energy modernisation. However, this legal status quo reduces landlords' motivation to invest in energy modernisation, given the prolonged investment payback period. Landlords in regions with less competitive rental markets additionally face risks associated with early contract terminations and potential dwelling vacancies.

Given the acute need, particularly in densely populated areas such as Berlin, to provide the population with rental housing at reasonable rents, and in light of Germany's climate protection goals, it is scarcely feasible to resolve the landlord-tenant dilemma solely through tenancy law reform, without forgoing energy modernisation measures. Therefore, seeking a solution beyond tenancy law within this thesis was essential.

The analysis of Swedish legislation has shown that the Swedish legislator does not guarantee any enforceable right for vulnerable groups to claim a specific level of energy efficiency. Tenancy law offers only the lowest acceptable standard of living. A major barrier to energy modernisation in Sweden is the rent control system, which is extensively

[3] Ibid.

[4] 'Wohnungsbestand wächst regional unterschiedlich'. 2022. 30 June 2022. https://www.statistik-berlin-brandenburg.de/143-2022.

criticised in the literature. Rent control diminishes the landlord's incentive to maintain and enhance dwellings within the tenancy. Inclusive rents contribute to split incentives for tenants, as they do not directly benefit from reduced heating costs resulting from energy modernisation, leading to rebound effects.

In searching for a solution to the landlord-tenant dilemma outside of tenancy law, public and private financial instruments were examined for their potential to mobilise investment in residential buildings' energy modernisation. KfW loans were ruled out as an effective solution since they are available only as supplementary financing to applicants' existing financial assets. They fail to address the landlord-tenant dilemma and are inaccessible to low-income tenants lacking the requisite financial assets. Landlords, on the other hand, are disinterested in investing in energy modernisation, even with grants, as they do not directly benefit from the energy savings. Reforming public law levies is exceedingly time-consuming and fraught with constitutional risks. The warm-rent model exacerbates usage-related split incentives by fully transferring heating costs. The partial warm rent model also fails to resolve the landlord-tenant dilemma, merely redistributing the costs of energy modernisation to the landlord. Furthermore, the current modernisation rent increase cap of 8% does not provide sufficient incentive for landlords to decarbonise dwellings. Information asymmetry means landlords, especially in less competitive rental markets like Spree-Neiße, are justifiably concerned about tenants moving out and leaving their properties vacant. Additionally, this approach does not address the lack of financial accessibility for landlords without available resources. Moreover, the cap renders renovation loans economically unviable due to the prolonged repayment period.

Sweden is well-known for financing energy transition projects through municipal and corporate green bonds, an instrument successfully employed by both large and small Swedish municipalities. In contrast, German municipalities have yet to adopt this instrument. This thesis investigated the applicability of this instrument to Germany and the conditions necessary for its implementation. EU, federal, and land-level laws were examined for legal barriers to the issuance of municipal green bonds. Generally, EU law does not forbid municipalities from issuing green bonds, although specific considerations, such as emissions by investment funds in which municipalities hold shares, or by entities acting as market producers, must be considered.

These considerations are not so significant as to render the issuance of green bonds by German municipalities uneconomical. Moreover, there is no precise federal constitutional debt brake in the German legal system that explicitly prohibits municipalities from issuing green bonds. However, legal uncertainties remain for municipalities. Legally, it would be beneficial to introduce clear regulations that unambiguously permit credit financing of the energy transition at all levels, including municipalities. Some *Länder* constitutional laws do have restrictions that impede the issuance of green bonds. The main challenge, though, lies in ordinary legislation, which restricts municipalities' ability to finance debt, with only limited exceptions that entail additional legal risks.

Two solutions within the current legal framework were proposed, considering these legal barriers: the issuance of green bonds on behalf of municipalities by legally autonomous entities and by private credit institutes. The compared regions, Berlin and the Spree-Neiße district, are governed by different Länder legislation. Berlin, as a city-*Land* and the capital of the Federal Republic of Germany, faces additional legal barriers to issuing green bonds compared to the municipalities in Spree-Neiße. However, the proposed solutions are applicable to both regions. Additionally, case law from the Constitutional Court of Berlin indicates that the debt brake does not apply to independent legal entities, providing further legal certainty for this *Land*.

From a non-legal perspective, the proposed financial mechanism is viable for both regions. It would enable private investment mobilisation not only in economically robust metropolitan areas like Berlin, which generally benefit from the energy transition and attract investors, but also in economically weaker regions such as the Spree-Neiße district. Municipalities facing high financial barriers could form associations for joint energy transition projects, financed by municipal green bonds under the proposed scheme.

References

'Wer profitiert vom Berliner Mietmarkt?' 2019. Tagesspiegel. 28 June 2019. https://interaktiv.tagesspiegel.de/lab/mieten-und-renditen/.

'Wohnungsbestand wächst regional unterschiedlich'. 2022. 30 June 2022. https://www.statistik-berlin-brandenburg.de/143-2022.

GPSR Compliance

The European Union's (EU) General Product Safety Regulation (GPSR) is a set of rules that requires consumer products to be safe and our obligations to ensure this.

If you have any concerns about our products, you can contact us on ProductSafety@springernature.com

In case Publisher is established outside the EU, the EU authorized representative is:

Springer Nature Customer Service Center GmbH
Europaplatz 3
69115 Heidelberg, Germany

Batch number: 10739812

Printed by Printforce, the Netherlands